TECHNOCRATS AND NUCLEAR POLITICS

Technocrats and Nuclear Politics
The Influence of Professional Experts in Policy-Making

ANDREW MASSEY
Department of Political Studies
Public Policy Research Unit
Queen Mary College
University of London

Avebury
Aldershot · Brookfield USA · Hong Kong · Singapore · Sydney

© Andrew Massey, 1988

All rights reserved. No part of this publication may be reproduced, stored in a retrieval system, or transmitted in any form or by any means, electronic, mechanical, or photocopying, recording, or otherwise without the prior permission of Gower Publishing Company Limited.

Published by

Avebury

Gower Publishing Company Limited,
Gower House, Croft Road, Aldershot,
Hants. GU11 3HR, England.

Gower Publishing Company,
Old Post Road, Brookfield, Vermont 05036
USA

ISBN 0566 05644 5

Printed and bound in Great Britain by
Athanaeum Press Ltd, Newcastle upon Tyne

Contents

Preface   vii

Chapter 1, Introduction   1

Chapter 2   The concept of the   20
            'Technical Professions'

Chapter 3   Science and Engineering   35

Chapter 4   The Professional Structures I   52

Chapter 5   The Professional Structures II   75

Chapter 6 The Reactor Decisions I	107
Chapter 7 The Reactor Decisions II	125
Chapter 8 The Evolution of British Nuclear Fuels Limited	156
Chapter 9 The Windscale Inquiry	172
Chapter 10 Conclusion	185
References and Bibliography	191

Preface

The book is concerned with analysing the role of technical experts in the development of Britain's civil nuclear energy policy. There is a specific focus upon scientists and engineers, their interaction with and influence upon the political system and the policy process. The major theme of the work is the proposition that civil nuclear initiatives flowed from the integration of technical professions within the bureaucratic hierarchies of the governmental and quasi-governmental organisations involved in the formulation and implementation of policy. A combination of organisational logic and professional motivation (informed by the ideology of "professionalism") acted as a spur towards the evolution of policies by the technocrats that were calculated to advance their goal of occupational autonomy.

There is an examination of the policy process which analyses the networks of influence, integral to the satisfactory implementation of initiatives within the unitary British political system. Specifically, the interactions between technocrats, civil servants, politicians and industrialists. The study argues that civil servants and politicians perform the role of policy arbiters, selecting from among the options constructed for them by the technocrats.

The first part of the book develops the concept of 'technical professionalism' and places it in the context of previous work in this interdisciplinary field (Chapters 1 to 3). Chapters 4 to 7 then focus on the technocrats' role in providing a spur (or dynamic) from their positions within the policy community's bureaucracies, that drives top-level policy decisions.

The book then examines these functions in more detail via the use of case studies. The two studies are concerned with the evolution of an independent fuel-cycle industry and the Windscale Inquiry. Ways in which professional ideologies and frames of reference have defined the limits of public policy debates and circumscribed political choice, whilst paradoxically resulting from strong political initiatives are examined in detail, providing an understanding of the issues involved in the contemporary nuclear debates in Britain.

It is notable that research conducted within the field covered by the social sciences often involves persuading people to answer questions designed to assess their motivations for doing, or not doing, certain things. To this extent any policy study is something of a "team" effort dependent upon the co-operation of disparate groups of people who may have little or no interest in the goals of the researcher, but who nevertheless condescend to grant interviews. To those men and women who kindly allowed me to quiz them during the course of this thesis I am profoundly grateful, as without their forbearance, honesty and courtesy, material intrinsic to the research would not have been available. The interviewees (of which there were over one hundred) included politicians, civil servants, lawyers, industrialists and others associated with nuclear power in Britain and Europe. Most people, however, stipulated a wish to remain anonymous, and although not named their contribution is gratefully acknowledged.

The work for this book began as a PhD thesis at the London School of Economics under the guidance of Dr. Patrick Dunleavy. His assistance and analysis were a formative influence upon parts of the research and I am grateful for his skill and kindness during my period at the School. Whilst at the LSE, I am also grateful for the financial support provided by a grant from the ESRC. Following a transfer to Queen Mary College, Professor Trevor Smith supervised the thesis and it is to him that I owe the greatest debt. His criticisms, advice and encouragement were everything hoped for from a supervisor and much of the credit for the completion of that and also this work falls to him. Following the work for the PhD, Professor Smith provided further assistance and advice in the complete rewriting and updating that eventually emerged as this book. Readers should be aware that an early draft of Chapter 8 appeared as a paper in <u>Politics</u>, volume 6 number 1.

Many thanks are due to Pat Bellotti for some of the typing and to Professor Mike Goldsmith of the University of Salford for logistical support and some enlightening debates. Most of all I owe a debt to my family for their support and forebearance.

All errors are, of course, my own.

Andrew Massey.

1 Introduction

THE PURPOSE AND GENERAL THEMES OF THE BOOK

The contemporary scene

For those practitioners and observers with an interest in the peaceful exploitation of nuclear power in Britain, 1986 was something of a watershed. Three events occurred that will have profound implications for the future of civil nuclear power in the United Kingdom; on Sunday, April 26th at 01.23, the Chernobyl nuclear power station in the Soviet Union suffered an explosion that released several tonnes of radioactive material to drift over Europe; that same month Mr Christopher Harding was released on a part-time basis by the Hanson Trust to begin work as Chairman of British Nuclear Fuels PLC (BNFL); and at the end of the year Sir Frank Layfield presented his report on the Sizewell 'B' Inquiry to Mr Peter Walker, the Energy Secretary, recommending an American-style Pressurised Water Reactor (PWR) fuelled electricity generating station be built in Suffolk. In terms of media coverage, and therefore public attention, the sensation of the accident at Chernobyl comfortably outstripped the other two. But the long-term effects of Harding's appointment and the Sizewell Report (so elegently written by Layfield) have a potential impact upon that same public that outweighs the possible consequences of the Ukrainian disaster. Taken together, the three events have irrevocably shifted the perspective of the nuclear debate in Britain.

From its inception the nuclear industry has sought to promote public confidence in itself and has taken great care to stress the inherent safety of its power stations and associated plants (cf.Gowing,1974;Williams 1980). The steady growth of anti-nuclear sentiment in much of Europe and America has eroded this confidence. It was compounded in Britain from the mid-1970s with a succession of accidents at the Sellafield nuclear complex, culminating in

1

prosecution and fines for BNFL and the official displeasure of the House of Commons Environment Committee, which was expressed in a March 1986 Report (Cmnd, 191-1). The following month operators at the Chernobyl nuclear power station were to exacerbate the international reputation of the industry even further when their bizarre experiments with the turbo-alternator (whilst the automatic safety system was switched-off) led to an excessive amount of heat being generated, part of the reactor went prompt-critical and the explosion lifted the pile-cap, releasing ten to twenty per cent of the reactor's radioactive iodine and ceasium over Europe (Atom, 360, p.6). For a while in Britain the weathermen were national heroes whenever they announced a favourable westerly wind. Over 135 thousand people were evacuated in the Soviet Union due to the contamination and throughout Europe millions of tonnes of food were rendered inedible; for example in Wales and the North West of England farmers were still prevented from slaughtering their sheep a year after the accident. The number of people who will eventually die of cancer in Europe as a result of the contaminated wind and rain might have to be numbered in their thousands (Atom, ibid, p.7).

Chernobyl, therefore, reinforced a growing unease about nuclear safety and a perception amongst sections of the British populace that the assurances of technical experts were not a reliable safeguard against such accidents. The United Kingdom Atomic Energy Authority's (UKAEA) house journal, Atom, caught the mood by regretting that:

> "as in the case of the Three Mile Island accident, what happened at Chernobyl again emphasised the importance of the human factor and of the inter-relationship between man and machine".(ibid, pp.7-8).

The author could also have added the accidents at Bhopal, Flixborough, the American Space programme, the legacies of DDT, Dioxine, and nitrates to the concatenation of factors illustrating the fallibility of human control over technology. Yet in Britain the constant reassurances from the nuclear industry show a refusal to acknowledge a potential for a major disaster at British nuclear power stations; Chernobyl highlighted woefully inadequate preparation of emergency plans should such an accident occur (Guardian, 6/6/86). Even before the Russian disaster Oppenshaw wrote:

> "Quite simply emergency plans are bad publicity for the nuclear industry because of the problem of explaining their existence for plant which is, of course, completely safe and does not need it !"(1986, p.284).

Chernobyl was the subject of intensive media analysis and speculation, but one important, albeit rarely noted factor, was seized upon by Hugo Young, "at least half the population," he wrote, " does not accept that the new radioactivity (from Chernobyl) is harmless." That is, there is a "crisis of credibility," in that people no longer believed the official reassurances (Guardian, 8/5/86). Clearly not only the professional status, but the veracity of technical experts was under question, a glaring contrast to the immediate post-war years of the inchoate nuclear industry when the "boffins" were held in esteem (Gowing, 1974; Williams, 1980).

In exposing the fragility of the consensus surrounding the professional standing and consequent influence of technocrats, Chernobyl served to focus upon an important theme of this book; the role and influence of technical experts in policy-making and the nature of their "professionalism". It is obvious their professional status is different to that of doctors and lawyers, and the question of safety has exposed a degree of partisanism rarely glimpsed in the law or medicine. These factors are explored in Chapter Two and Three below, preparing the way for the analysis of nuclear politics from the perspective of professional power.

In the analysis of nuclear politics in Britain, technical arguments must take their turn amongst the ebb and flow of commercial, military, and industrial considerations pondered by the policy-makers. The debate between the advocates of American-style PWRs and the British designed gas-cooled reactors rumbled on over more than two decades, until the commercial and industrial benefits attributed to the PWR finally combined with the mediocre record of the British AGRs to persuade the Conservative Government in 1979 to announce the intention of building a PWR at Sizewell. Following the aeon of the 340 day investigation carried out at the Public Iquiry, the Inspector's Report interpreted the evidence (of 195 witnesses, two-hundred written proofs of evidence, five-hundred addenda, fourteen written statements, over four thousand supporting documents, four thousand letters, and one-hundred-and-twelve further written submissions (Layfield, 1987, Ch1, p.4)) to be weighted in favour of the Central Electricity Generating Board's (CEGB) application to build a PWR at Sizewell. Layfield found in favour of the PWR, with some minor reservations, on grounds of economic efficiency, environmental impact, and considerations of safety (1987, Ch.2, General Conclusions,Ch.108). This recommendation was the one sought by the CEGB, the private manufacturers and the Government; Peter Walker wasted little time in giving formal permission to proceed, and after a perfunctory Parliamentary debate this permission was granted in March 1987.

Walker's decision marked the end of the British monopoly within the domestic nuclear market, it linked the United Kingdom's civil nuclear programme to the nuclear markets of the world which are dominated by the PWR, bowing to CEGB and commercial pressures for that linkage. So confident was the Board of the eventual decision that before the Inquiry was even half-way through the CEGB had ordered about £100 million worth of PWR hardware (Openshaw, 1986, p.68). Constitutionally such inquiries serve to "inform the minister's mind" and not as a forum for popular participation in decision-making, but such official confidence as that shown by the CEGB can only further undermine the faith of those excluded from the closed world of policy-making in the justice and responsiveness of the political system in Britain. Clearly the Inquiry was not rigged and the nuclear industry awaited Layfield's Report with same trepidation, whilst severely criticising the Inquiry for its drain on their resources in time and manpower. But the entire process from Howell's announcement in 1979 to Walker's 1987 decision was obviously weighted in favour of the PWR proponents; a point noted by Layfield at the start of the Inquiry and in his Report when he referred to the lack of resources at the command of the objectors (Ch.2, p.8). The decision to build PWRs in Britain reversed thirty years of stategic and commercial Government policy which was designed to protect British technology ; it is ironic that a commercial rationale was used to underpin this move. Chapters Four to Seven explore the interaction between technocrats and policy-makers in detail, analysing the influence of technical, commercial, and military considerations that coalesced in the 1987 decision.

Oppenshaw was amongst those who lacked a conviction for the PWR and was scathing in his condemnation of that option, arguing that:
"the US nuclear industry plans to survive the current depression partly by supplying their obsolete (in terms of US criteria) plant to countries stupid enough to think they are a good buy."(1986, p.320).

An integral part of the British decision in favour of the PWRs was the obsession with strategic independence of fuel supplies, combined with the increasing emphasis on commercial efficiency. This led inexorably to the PWR with the CEGB bluntly arguing that in effect "either you accept the proposed developments or you face power cuts" (ibid, p.47), and opponents to nuclear power are viewed by the industry and the Government as "misguided or in favour of a different, non-industrial life-style" (ibid). A programme of PWRs in Britain,then,marks a break with British-designed technology and a greater emphasis (or perhaps re-emphasis) on market-orientated commercial efficiency. This is a thread that runs through much of the analysis of nuclear politics, not least in the field of the nuclear fuel-cycle, described in Chapters 8 and 9. The appointment of Mr Christopher Harding to the Chairmanship of Britain's nuclear fuel industry is significant in that it marks the end of technocratic control and the triumph of the concept of commercial dominance.

BNFL's new internal slogan may well be,"there is no industry unless we have a safe industry," (or variations thereon) but the unspoken slogan is that there is, "no industry without profit". It is the contention of this book that the engineers of the nuclear fuel industry, acting through a combination of professional and organisational motivations, used the logic of the competitive market to secure autonomy for themselves within the nuclear policy system. With the triumph of the theories of neo-classical liberalism, however, that manoeuvre was used by the post-1979 Conservative Government to impose a commercial profit-orientated regime upon the technocrats, once again subverting their autonomy to Government aims. The new regime was personified in the appointment of Mr Christopher Harding who,personally charming and urbane,brought with him from the Hanson Trust the mores of the private corporate sector and the determination to inculcate them into BNFL. In short, the quiet transfer of power from Con Allday to Mr Harding which was lost amid the clamour of adverse reports on radioactive leaks at Sellafield,Parliamentary opprobrium and judicial fines, represented a perceptable shift in the direction of privatisation.

This move was a logical progression of the company's explicit commitment to commercial success and the sections of the 1971 Act, which established BNFL, making provision for private participation. It also came as an inevitable extension of the importation into Whitehall and the public sector of private sector managerial techniques, such things as the Raynor scrutinies, MINIS, and the Financial Management Initiative (Likierman, 1982, pp127-142). Managerial accountablity, profit centres and performance indicators were instilled into the structures of the public sector from about 1980 onwards and in 1986 BNFL adopted these techniques to an even greater extent than it had already done so. The company was reorganised, sweeping away the technocratic concern for function and concentrating upon profit-centres. Specifically these were to be fuel fabrication, enrichment, and reprocessing (Annual Reports and Accounts,1985-86). With this came something of a devolution of managerial power, providing "the person in-charge of a function total responsibility and total accountability," as one senior manager put it. It could be argued that despite the special ire that environmentalist groups reserve for the company (resulting in an unprecedented advertising campaign by BNFL in 1986 and 1987 to clean-up its image) the nuclear fuel industry could lead other parts of the nuclear conglomerate into the private sector. Certainly similar reorganisations and thoughts have afflicted the UKAEA and CEGB in recent years (UKAEA Annual Reports, 1980-1981 to 1985-1986; CEGB Annual Reports, 1981-1982 to 1986-1987; Kemp, 1986, pp335-341).

Some General Themes

Consideration of these issues has led to the essential purpose of this book, which is to examine the motivations and varied perceptions which laid the foundation for Britain's contemporary civil nuclear industry. Of particular concern has been an attempt to analyse the role and influence of technical experts in shaping the decisions of their own industry and the implementation of those decisions. Because the focus underlying this study of the policy-system of civil nuclear power is that of the political scientist, it follows that one of the central goals of the analysis is to use this exploration to elucidate the operation of Public Administration in Britain, specifically the methods employed in the processes of decision-making in an area renowned for the high technical content of its options. In pursuing this aim the detailed analysis of the role and function of the technical experts employed by the various organisations involved in the nuclear decisions is a crucial part of the study and is undertaken in the chapters which follow.

As a complex and highly expensive technology, the nuclear programme has become the concern of a potpourri of state and private organisations: Westminster, Whitehall, local authorities, state-owned industries, quasi-governmental agencies (QGAs), private manufacturing industry, the trades unions and conservationist interest groups, have all become involved with the process of decision-making. The type and scale of the United Kingdom's commitment to nuclear energy as a means of generating electricity is a result of the interactions inherent to the organisations involved in the policy process pursuing the implementation of their goals, and the necessary compromises (or not) therein. An aim of this study is to seek to explain the formulation of nuclear policy in Britain's centralised and unitary system of government through an illustration of the manner in which the entities involved interact with each other. As indicated above, one of the mediums of interaction increasingly important in the Political Science literature is that of the professional experts; of professionalism. In a perspicacious article, Dunleavy argued that the implications for policy making of professionalism:

> '...have been conceived almost exclusively in terms of studying state responses to attempts to professionalise occupations, rather than asking about the implications of a professionalised state apparatus for substantive policy formulation' (1981, p.4).

A recurring theme throughout the book is the development of the interaction of the technical and political worldviews and the tensions that frequently result from this. Particular attention is given to the problems the political leadership encounter in a technically-dominated area in enforcing the political functions of control and accountability, in both the formulation and implementation of policy. In essence this leads to the examination of three basic propositions. The first of these is that it has been argued that the professions and professionalising occupations form an alternative communications system. That is, professional contacts and activities 'effectively unify policy development across formal agency boundaries' (Dunleavy, 1981, p.10). The professionalised occupations and the agencies they predominate in will come to display a 'national-level ideological system' (ibid). The implications of this are clear, the major one being that the political leadership, faced with a united professional opinion on highly technical matters, will find its authority subverted by the need to defer to the expert views of its advisers. On a more routine level it would lead to the homogenising of policy options as presented by the technocrats. One of the arguments of this book is that the history of civil nuclear power tends to undermine this stance.

The second general proposition is a progression of this and argues that a professionalised policy-system leads to a reduction in the political content of decision-making. It can be seen that if there is a diminution of the policy-making role of the political leadership and an arrogation of that role by the technocrats, a logical concomitant of technical decision-making is a reduction of the political content in that process. Certainly a variety of researchers have seen this as a goal of those able to attain it (Larson, 1977 pp.38-191; Dunleavy, 1981, ibid; Wilding, 1982, pp.13-78; Laffin, 1986).

The third general proposition tested by this book is that because the policy system is imbued with a high technical content, and because this leads to an important role for the technocrats in explaining policy options to the political leadership, it seems logical to suggest that the decisions opted for in the debates on nuclear power were a result of the implementation of professional preference. Politicians and civil servants, it could be argued, were guided by the wishes of the technical experts; opinions in turn guided by the exercise of rational professional judgement. The resulting policies were therefore professional and not political choices, the political elite being reduced, therefore, to the role of policy-arbiters and not policy-makers, their primary function being to select from amongst a limited number of technocratically formulated policy options and that selection itself deeply reliant upon technocratic advice.

These propositions are suggested by the existing literature and are tested throughout this study, the literature pertaining to the professions being reviewed in the following chapter, although an introduction to it is provided below (this chapter). Whilst a review of the literature allowed these three general propositions to be formulated, the method of testing them presented itself as a more arduous challenge, the secretive and hidden nature of the policy-process in British government seemingly providing an effective shield with which to protect itself from the prying eyes of the researcher. Many essential documents concerned with the decisions will remain restricted by the Official Secrets Act for years to come and the minutes of meetings held by the private sector and state-owned industries will remain similarly embargoed due to the demands of commercial secrecy. Those that have been released have been sanitised for public perusal, those parts revealing the original (and continuing) overlap between the military and civil uses of nuclear power remaining, for the most part, unread by all but the Whitehall 'sifter' and those privileged to be accredited with the requisite security clearance, an honour accorded to few academics. The difficulties encountered with the written word are repeated with the spoken version. The tradition of anonymity and secrecy within the protective cloak of collective and ministerial responsibility, is ingrained; the ethos of Whitehall and Westminster is such that access to individuals is rarely granted and, on the few occasions when it is, most quotes have to remain 'unattributable'.

Relationship to Previous Work

This research lies at the intersection of several fields of study, most notably the works concerned with nuclear power, professionalism and public administration/policy analysis. In a very real sense each chapter contains its own literature scan as the scope of the subject is beyond containment at any one juncture. It is necessary at this point, however, to provide an introduction to the different approaches found in the literature of the subjects in order to locate this book.

Nuclear Power

There are several approaches to the study of nuclear power. The first source for any student must be the official history, provided by Margaret Gowing, that in two volumes charts the development of the technology from the troubled late 1930s until the potential for civil exploitation became a reality in the early 1950s (1964 and 1974).

Using official sources, Professor Gowing provides an illuminating insight into the political rivalries and manoeuvrings that led to the establishment of the Atomic Energy Authority, the exploding of the Anglo-American and British atomic bombs, and the building of the Calder Hall and Chapel Cross power stations. When this is combined with the accounts and official biographies (and autobiographies) of those involved in the early years (Cockcroft, 1950; Hinton, 1958) what seems to be the official ideology can be pieced together. The accepted and explicitly-stated aims of the policy-makers are revealed and can be contrasted with the achievements in the implementation of the policies in order to provide the basis for an evaluation. This approach is, therefore, helpful in that it allows a measurement of achievement: its weakness, however, is that frequently the stated goals provide a cover for implicit or unrecognised aims. From a political science stance, its main weakness is that it is not primarily concerned with politics per se, except as they intrude upon the main themes of the study which is technical development. Understanding of the policy-system is obscured and therefore must be gleaned from among the other considerations.

Another school of thought is provided by those who are in general agreement with the expansion of nuclear power, but criticise the manner in which this has been pursued in Britain. One of the severest of these critics is Burn, who has argued forcefully for a greater involvement of the private sector unfettered by official bodies and their concomitant regulations (1967 and 1978). His championing of the free market reflects a belief in the liberal market economy which, he has argued, would have led to greater efficiency and followed the American example in exposing the deficiencies of the British programmes before they became severe technical and commercial liabilities.

Whilst stating that, "it is now abundantly clear that the future source of electricity in many countries is seen as being nuclear," and that, "dependency on nuclear generation will gradually increase," Openshaw is sharply critical of the methodology employed for the siting of nuclear power stations in Britain (1986,p.2). In combining the expertise of the geographer with advanced statistical analysis, he has written a powerful critique from the perspective of the case for a safer siting policy that takes full cognicence of population densities and prevalent environmental factors; an argument possessing an added emphasis since the events of Chernobyl and the continuing debate over cancer clusters in the population near nuclear plants (c.f. J.Urquart, letter to The Guardian, 13/1/87). Openshaw seeks a wider input into siting policy to give a greater participation to experts other than CEGB engineers, and indeed it is his goal to open-up the policy-making process to a wider public per se. He argues that:

> "it is hardly satisfactory that enormously expensive investment decisions should be made in secret without any real public debate and with the minimum of opportunities for effective public participation and inputs" (1986, p.11).

The decision-making process is, he argues, "immune to either local objections, or national political control, or anything outside the narrow view of the planning engineer and the Board member" (ibid). Whilst the former points are certainly born-out by much of the research carried out for this study, the latter quote does not stand up to close scrutiny within the context of the British political system; the substantive part of this book argues the supremacy of "politics" over "technology" in Britain

Sweet offers an economic critique, but has propounded the opinion that the nuclear power programme is wasteful and inefficient per se (1982). The thrust of his assault has been aimed at the Central Electricity Generating Board which, he argued, has hidden the real costs of nuclear power through a mixture of Byzantine accounting practice and wilful obfuscation in order to protect and promote the nuclear sector. He argued:

> 'The only advice to the uninitiated in the art of CEGB accounting is never to accept a single figure evidence that nuclear power is cheaper than anything else. Such figures are always misleading, if not downright wrong'.(1982, p.33)

The result has been a massive overemphasis on nuclear power within the Electricity Supply Industry (ESI) and its technical support, according to Sweet. The result, he maintains, is that Britain:

> '...is lagging behind in new technologies - Combined Heat and Power in particular - for no other reason than that we have a centralised electricity authority which uses its monopoly to promote nuclear power to the virtual exclusion of better placed alternative technologies' (ibid. p.27).

The literature therefore contains something of a bifurcation in the discussions pertaining to the efficient utilisation of resources. One branch in favour of an expansion of nuclear power but critical of the method, the other branch critical of the whole concept, seeking to challenge the policy-makers on the basis of their own figures and claiming to have found them wanting.

A sharp contrast is contained in the approach taken by those fervently in favour of a massive expansion of nuclear power. Of these Hoyle (1977), and Greenhalgh (1980) are among the most prominent; Hoyle going so far as to cast anti-nuclear campaigners in the role of Soviet dupes. The central theme of this approach is the strategic one concerning the rapid depletion of fossil fuels available to the West and the vulnerability of countries reliant upon Middle Eastern or Soviet reserves. An expansion of nuclear capacity is seen as the only alternative to a creeping immiseration of society as coal, oil and gas run down.

A more thoughtful analysis is that adopted by Roger Williams (1980). In what has become a classic account, Williams has traced the historical development of nuclear technology and the integration of the technical options with the political perspectives of those responsible for making the decisions. He argued from the position that:

> 'What was really wrong was that the decisions were not, for the most part, framed with a view to enhancing policy flexibility later: they were insular and primarily designed to maximise gains rather than minimise losses' (1980, p.321).

This he blames upon the structure of the policy-system that, despite its plethora of experts, left the political (and therefore legitimate) decision-makers uninformed. He argued that in Britain's nuclear development:

> '...government mostly amounted to ratification, indifference or bewilderment...Faced with matters which they could not really be expected to understand, decision-makers...on grounds of commercial security or constitutional propriety (had) the arguments closed up and decisions made on an inadequate basis of true knowledge'. (ibid, p.328).

Williams's perspective, therefore, is essentially one that viewed the nuclear decisions as a result of inadequate policy-making procedures, in particular the propensity to exclude popular participation from the process and indulge in what he termed 'private decision-making'. His remedy involved a more open and informed style of policy-making in which politicians and public alike were fully briefed as to the options available.

Finally, there is the approach that embraces outright rejection of nuclear power. This can be sub-divided into rejection on moral grounds, environmental and safety grounds and a political/ideological critique. There is, of course, some overlap between these divisions and the economic critique by Sweet (and those who adopt his approach) as outlined above. The central theme of those who advocate a moral reproach against expanding nuclear power is the perceived link between the civil and military uses of atomic energy and a belief that an expansion of the former facilitates a similar increase in the potential use of the latter. Hesketh [1], Pharabod (1983), Rotblat and Brown (1983) are among a group of informed nuclear experts who have viewed an increase in the number of civil nuclear reactors as an aid to the military programmes of those states in possession of atomic weapons. They have argued that since a concomitant of the daily workings of a nuclear power station is the production of plutonium, the raw material for nuclear weapons, and since there are strong grounds for believing that at least some of that metal is passed onto the military, it follows that an expansion of the civil aspect can cause a similar expansion of the military use. Environmentalists and safety objectors such as Friends of the Earth [2] contend that the long time-scale involved before elements like plutonium become safe for human contact (several tens of thousands of years in some cases), and the horrific potentials of an accident (or terrorist bombs), make the use of nuclear power tantamount to criminal folly. Society, they argue, owes it to future generations not to leave a legacy of deadly poisons which may never be cleaned up. Their alternative is safe, or 'soft', energy such as wind, wave and solar power (Conroy, 1978).

The political/ideological opposition is more complex and by its nature more ad hoc. Its proponents such as Rozack, Marcuse, Habermas [3], and Wynne (1982) adopt quite different analytical stances from one another. Yet there is a basis of similarity in their approaches in the sense that the increase of state intervention into civil society demanded by the peculiar nature of atomic power and its attendant risks and potential, leads to a need for the state to legitimate that intervention. In mature capitalist societies, the intervention takes the form of an implicit support for Corporate Capitalism and the legitimation is achieved through the depoliticisation of social and political issues and their presentation as straight-forward commercial and/or technical decisions to be taken by those technically expert to do so. Thus the professionally qualified become a party to the ritualisation of issues 'as a discovery and not as a social choice' (Wynne, 1980, p.169). This leads to the contention that:

> "...the empiricist framework thus systematically deflects attention from the kernel of the issue, which is political control and legitimation. For modern technology, embodying large-scale disciplined social relationships, possible material consequences are dwarfed by unforeseeable changes in social being" (ibid. p.164).

This approach, therefore, is the opposite of that employed by those attached to the liberal ideology in that it views the nuclear programme as a manifestation that embodies many of the more extreme aspects of Corporate Capitalism to which this approach is opposed. It is a perspective which attempts to reveal the structure of the policy-system as an aid to a conspiracy that legitimises authoritarian decision-making, ideology (either liberal ideology or scientific/professional ideology) being used to strip political and social decisions of the subjective reality and have them made objectively by experts as problems of technical choice.

The existing work on the study of the political implications of nuclear power falls broadly into the somewhat overlapping categories outlined above. In the chapters which follow use is made of all these and the information contained therein, but it is the latter group which provide the most useful clue as to the nature of the decision-making process. This is not to accept the ideological underpinnings of the works of Wynne or Habermas (nor to completely reject it): rather it is to utilise the insights they can provide into individual and organisational motivations. It may be that such insights attribute the wrong motivations to individuals acting on their own or collectively through organisations, but their emphasis on analysing the role of ideology and in particular the ideology (or ideologies) found within the structural constructs of modern liberal-democratic states provides an important link both with the literature concerned with the study of professionalism and through that to some of the more relevant public policy/ administration literature. It is from Chapter 4 to Chapter 9 that these approaches are combined with a reformulated view of technical professionalism and applied to the nuclear decisions of the past two-and-a-half decades, allowing the synthesis of a view of those decisions which attempts to aid the understanding of the nature of public administration when applied to major technical projects.

Professionalism

As with nuclear power, and most complex subjects for that matter, the analysis of professionalism has attracted a fairly wide range of approaches. The major discussion of these schools is situated in the following chapter where it is attempted to construct an explanatory model of the technical professions which is then applied to the engineering and scientific occupations. Traditionally observers of professional behaviour have adopted a 'trait' theory to attempt to understand this occupational phenomenon, a phenomenon for which there is still no widely accepted definition. Trait theory simply attempts to list all of the characteristics or 'traits' that an occupation needs to display in order to qualify for the coveted status of being accepted as a profession. Theorists belonging to this school have a venerable history and include Lewis and Maude (1952), Millerson (1964) and Greenwood (1965). A critique of this stance, Wilding (1982), notes the major weakness as being that the writers themselves are often the captive of the professions' own ideology and therefore unable to distinguish the role of the professions from a social theory viewpoint. As a result, trait theory is usually innately conservative and useful as a vehicle for the professions and professionalising occupations to use to advance their own goals, examples being found in the Monopolies Commission Report (1970) and the Finniston Report (1980), both of which accepted the trait (and therefore professional) descriptions almost without question.

A second approach is that of the 'functionalists' who see professionalism as having a functional relevance either for society as a whole or in the relationship between professions and their clients. Advocates of this analysis include Barber (Johnson, 1972 pp.33-39) and Talcott Parsons (1954 and 1967). Johnson views the functionalists as possessing many similarities with trait theory and therefore it can be argued to share several of the defects (1972). Its main weakness, however, is that it adopts a consensual view of society that does not conform to the reality of its conflictual nature (ibid, pp.32-38). This, therefore, 'effectively eliminates from consideration the consequences of power relationships' both within society and the professions themselves, an omission that seriously damages the functionalists' ability to understand the social reality they seek to describe (Johnson, 1972, p.36). It certainly hinders a study of ideological and organisational motivations as a catalyst for policy change.

Durkheim saw the professions as a precondition for consensus in society, forming moral communities based upon occupational membership [4]. Tawney also looked for the enlargement of professionalism, seeing it as a force capable of subjugating individualism to the needs of the community (ibid). Whilst Carr-Saunders and Wilson pursued this theme (ibid), Weber had taken the view that professionalism was linked to bureaucracy and both were linked to the increasing rationalising of society (ibid p.15). This latter theme is one that has links with the political-rationality view seen in the above section on nuclear power (although nowhere near as developed) and indeed was pursued by C. Wright-Mills (ibid). It can be seen, then, that there are two broad themes which can be identified in the literature: that professions are moral defenders of the rights of the members of society; or that they are the allies of an increase in bureaucratic social control.

More recent writers have taken a somewhat different view, Illich for example recasting the theme of bureaucratisation to cast the professions in the role of disabling the individual members of society who come within their ambit for selfish reasons connected with the advancement of professional power and autonomy (1977). The professions, he argues, disable people by taking from them their right and in some cases their desire, to make everyday decisions connected with their lives, leaving them reliant upon the professions and thereby enhancing professional power. Freidson (1977) followed this particular line of argument although he has added that the main function of professionalism is the pursuit of privilege. Certainly there is little in this that Johnson would disagree with (1972), although he has added the rider that privilege and autonomy may well be the ultimate professional goals; but paradoxically the cost of attaining such a control over the occupational sphere has been an acceptance of the absolute right of the state to define the boundaries of that autonomy. As a result, Johnson argues, professionalism is a peculiar form of control over those occupations seeking professional status and in particular those with the potential power to attain it.

Larson (1977) and then Dunleavy (1979; 1981) have pursued a different tack (one with similarities to that of Wilding (1982)), in linking professionalism firmly to the ideology of mature liberal-democratic countries, in particular those prevailing norms found amid the ethos of large corporations in mature Capitalism. Larson contends that the possession of 'scarce knowledge and skills is the principle basis of modern professionalism' (1977, p.136), but the main function of ideology in capitalist society is to conceal the existence of class and the basic structure of class exploitation (ibid, p.156); there are clear links here with the approaches of Habermas and Wynne in the preceding section. Larson is convinced that professionalism plays an important role in legitimating the aims and attendant structures of modern liberal-democratic societies and that it does this through propagating the empiricist ethic. That is, by accentuating the technical and rational nature of professional expertise and opinion, professionalism encourages a belief in expert decision-making and a reliance upon expert opinion as the basis of policy, thereby facilitating the depoliticisation of political and social issues. She argues that:

> 'The ideological apparatus of the state standardises the general structures of the dominant ideology. Respect for science, for knowledge, for the professionals' 'construction of reality'

grows among masses of people who shall never see a university. This respect is embodied in the social hierarchy.' (op.cit. p.242).

This important analysis is expanded in more detail in the following chapter.

Dunleavy has built upon the Larson hypothesis to expound the theory of 'ideological corporatism' to explain the role of professional groups in society (1981). Specifically ideological corporatism is defined by him to be:

> '...an approach for analysing intersectoral and inter-governmental flows of ideas, knowledge and techniques. Rather than focusing on a specifically political process of elite interaction and integration, with groups and formal organisations as key sources of initiative and the policy process seen as producing a bargained compromise, ideological corporatism is centrally concerned with policy systems as communities' (Ibid, p.7).

In these communities there is an ideological unifying role played by the professions and 'the central focus' found in these communities and 'most systems of ideological corporatism in advanced industrial countries is the profession' (ibid, p.8). There are similarities here with the work of Wynne in particular, and this link plays a part in the attempt at a model of technical-professionalism located in the next chapter and used to analyse the role of the technocrats in the policy-making process in the chapters which follow. In a sense, the substantive part of this thesis is seeking to test the applicability of this approach to the policy community found in the nuclear sector. There are similarities too, with the literature on policy-making and administration relevant to this study, and also some overlap with that literature through the work of Dunleavy.

Public Policy and Administration

The literature concerned with public administration and policy analysis is vast and disparate and is constantly growing even larger and diverse. Such is the scale of this literature that it is perhaps easier to explain what is not relevant to this study before introducing what is. At first glance it appears that much of the literature on local authorities may be rejected as irrelevant. Except in a perfunctory way there is little contained within this research that is applicable to local government of central/local relations (the Sizewell and Windscale Inquiries being the exception). The work of Jones and Stewart (1982), though, is a reminder that the centre of government is far from the unitary and centralised Leviathan that it appears to be at first sight. Indeed, it is a somewhat federal collectivity of warring factions competing for resources and status. More importantly for this study, Jones's comments regarding quasi government are worth consideration (1982, p.924). Unlike much earlier work which was imbued with a desire to define and count the myriad organisations covered by the umbrella term 'quasi-government' (Hood, 1979; Holland, 1979), Jones cut to the heart of the matter when he

noted that these organisations confer what he termed excessive
patronage on the ministers and civil servants responsible for their
establishment and for appointing people to them (op.cit., p.924).
This, as Greenwood and Wilson (1984, ch.10) point out (following
Chester, 1979), when linked to the large increase in the growth of
such bodies since the establishment of the Welfare State has led to
doubts being expressed about the principle of accountability via
Parliament to the electorate. Jones's solution is predictable; he
believes such decentralisation should be to directly-elected local
government and not to quangos, in order to re-establish accountable
government (op.cit. p.924). Thus, through the literature on
central-local relations, a point germane to this study is discovered,
namely a consideration of the role and accountability of
public-sector organisations, whether publicly owned industries, such
as the Electricity Supply Industry, or quasi-governmental structures
like the Atomic Energy Authority, hived out of a government
department. The position of these technocratic organisations within
the political system is a major part of this work and is, therefore,
repeatedly studied throughout.

Dunleavy's arguments regarding ideological corporatism, as outlined
above, clearly have a relevance when viewing the intersectoral
cleavages across the political system and the public/private divide;
and it is the study of the role of the professional technocrats which
provides the link across these divisions as they inhabit
organisations found in all sectors. In his call for a radical
approach to public administration (1982, pp.215-224) Dunleavy
criticises what he sees as the pre-eminent focus in the literature
upon 'the institutions, organisational structures and
decision/implementation processes of government' (ibid, p.215). He
calls instead for a recognition of the changes in public
administration in terms of their 'integral connections with broad
social conflicts' and to understand that the 'social consequences of
government procedures are not accidental by-products but integral
elements in administrative change' (p.217). But it is in his
argument for recognition and the reintroduction of functional
explanations (p.219) in order to analyse administrative processes
within the broader social and political system that Dunleavy falls
foul of other theorists who tend towards a radical analysis (if not a
radical approach to public administration). O'Leary criticises the
use of functionalism by pointing out that apart from many other
weaknesses (see the section on professionalism above) 'there is no
necessary connection between functionalist explanation and the
ability to theorise connections between administrative changes and
social conflict' (1985, pp.345-351). O'Leary's critique also runs
deeper, warning against epistemic and methodological closure of the
discipline, a threat he perceives to be inherent in this approach.

Yet such a closure is not feared by others: indeed Dunleavy
cites Saunders's dualist model of the state as an example of a
possible move towards the radical approach (Saunders, 1981,1982;
Dunleavy, 1982, pp.220-221). There are other criticisms, however,
Gray and Jenkins arguing that it is difficult to accept Dunleavy's
view of the state as an 'overarching unity' or the 'vision of
national government dominated by the interests of capital (1985,
p.27) that is implicit in Saunders's model. They do accept, though,

'the danger of examining administration in isolation from external factors', particularly when many of these have 'contributed to its evolution' and 'influence its operation' (ibid). In attempting to construct a theoretical model of administrative politics, Gray and Jenkins are struck by an organisational approach, claiming that administrative behaviour:

> '...takes place within an organisational context and a great deal of this behaviour is political in that it reflects particular struggles to achieve a variety of ends' (op.cit., p.46).

As they point out, this develops 'a perspective strongly argued by Self' who advocated the need to establish a political analysis of public organisations, pointing out that:

> 'it is worth remembering that administrative politics is a much more universal phenomenon than the democratic politics of pressure groups and parties' (1977, p.6).

Clearly, therefore, an understanding of the organisational motivations of administrators in the governmental, quasi-governmental and private sectors is important for explaining the reasoning behind many of the stances adopted by those involved in the decisions studied in this book.

Such a consideration does not, however, extend to a detailed review of the predominantly American Public Choice School of writers like Downs (1967), Tulloch (1965) and Niskanen (1971) and their pessimistic belief that officials will always prefer their own self-interest to that of the public. Although it must be noted that, as becomes evident in Chapters 5 and 6 concerned with the organisational and professional motivations of AEA technocrats, the professional ideology does tend to provide a cover and excuse for such behaviour. Self may well dismiss such theories as simply crude (1977, p.11), but even he has admitted to their plausibility 'up to a point', and it was this limited plausibility that helped spur Dunleavy to construct his radical approach (1982, p.215). Thus, although the following chapters do not espouse the Public Choice message, it must be seen that at least some organisational/-professional motivation can be explained by this approach. Essentially, however, this study tends to follow the analysis of Gray and Jenkins and their alternative school of bureaucratic politics, although with some additions. In following Wade (1979) and Goodin (1982), they expand on Self's dismissal of Public Choice, to show that the approach's neglect of administrative history and disregard for the differences found between administrative units, within administrative units, and the roles of different organisations, negates much of the validity of the Public Choice argument (1985, pp.60-62). What is necessary, they argue (following Pitt and Smith, 1981), is a 'model that recognises the character and pattern of the differentiation in administrative organisation'; to this they add a political dimension (p.71). The analyst must be aware of the 'internal capacity of departments to pursue a variety of goals, both within and between administrative structures' (ibid). That is also the standpoint of this study, with the added premise that an inherent factor determining the goals of the technocrats in the quasi-governmental sector is professional motivation and that it is this which is responsible for the presentation of policy options.

In accepting the arguments of Gray and Jenkins in recognising the differentiation inherent to organisations, this book also tends to emulate their, and others', recent studies of governmental and quasi-governmental groups from an organisational perspective (Pitt and Smith, 1981; Hood and Dunsire, 1982). This is a form of analysis elements of which can also be found in Heclo and Wildavsky (1974), Pliatsky (1982), Zifcack (1984 and Ashford (1981a,b). It must be emphasised that although these works share similarities of approach in recognising the complexity of the administrative apparatus, they also contain considerable differences in their perspectives. One of the more salient points to be extracted from this literature, and one of particular importance in the analysis of later Chapters is that although the public and private sectors have many common features, particularly now that public organisations are tending to emulate private industry more than ever, 'certain values provide a series of distinct constraints to the management of the public sector bodies' (Pitt and Smith, 1984, p.5). The role of the professions in transmitting those values is, therefore, of great concern in the following chapters and essential to the testing of the general propositions outlined above.

It should be clear from the above discussion that the classical literature regarding the sub-discipline of policy analysis bears little relevance to the view of organisational/ professional politics utilised in the analysis of the following chapters. Certainly the rational synoptic techniques of Simon (1957) and the incrementalism of Lindblom (1959) do not appear to aid analysis. Indeed the recurring problem of deciding what is and is not incremental (Burch and Wood, 1983) is illustrated throughout the substantive chapters below, in that profoundly important decisions (which could not possibly be described as 'increments') were subject to a variety of influences, some of which were incremental in nature, but many of which were not. As to 'rational' choices, the decision-making process sought always to appear rational (at least those involved in it wished to present it thus) and the points raised in the sections above show that much of the literature is aware of the importance of the maintenance of a semblance of technical rationality as a cover for political choice which is usually based upon ideological a priori and parochial motivations. Even as an ideal-type such a model lacks credibility, as Self demonstrates, in that it is itself irrational in its neglect of political factors (1977, p.51). Recent literature has, of course, recognised this and taken the work of the classical writers and others such as Vickers (1965) and Etzioni (1967), to distil it with the research of those concerned to understand the nature of administrative politics in an attempt to construct a policy analysis 'for the real world' (Hogwood and Gunn, 1985). This policy-study certainly utilises in the following chapters (and attempts to develop) an analysis based upon an awareness of the politics of organisations informed by professional (and other) motivations.

As noted above, this introduction to the literature is informed and developed by the research of the succeeding chapters. In Chapter 2, the literature on "professionalism" is re-examined and an attempt is made to formulate a model of technical professionalism in order that successive sections of the thesis have an analytical base from which to proceed. The concept of "technical professionalism", or "technocracy", is the perspective from which the role of scientists and engineers are viewed and as such it is logical to begin with its definition. From this construction initial (and tentative) proposals are suggested with regard to the possible consequences upon policy, in both its analysis and practical effects.

Chapter 3 moves off from this initial general position to examine more closely the scientific and engineering occupations and to evaluate the extent of the concept of professional technocrats there. Science and engineering are the focus because as the dominant technical occupations within the nuclear power policy system their adherence to professional goals is a central theme of this study. In mapping the structure of these occupations and assessing their degree of professionalisation, the ground is prepared for an application of the theoretical implications of professionalism to the nuclear policy process proper. Specifically, an attempt to ascertain whether technocrats are responsible for the policies adopted by the United Kingdom in the nuclear sphere since the 1950s and the influence of professional motivation in those choices.

Chapters 4 and 5, therefore, serve as a lead into the body of the study in that they continue the focus upon scientists and engineers, but concentrate upon the structures in which they are located in the nuclear policy system, and the technocratic propensities of those organisations as a result of the position of the technical experts. It is in these and the succeeding two chapters (6 and 7) that the parameters of the research are simultaneously narrowed and expanded. Narrowed in the sense that there is a specific analysis of technocratic influence within the policy system and the decisions it took at different times regarding nuclear reactors; expanded in that such a study necessarily involves an examination of broader political and organisational perspectives. These four chapters provide the heart of the thesis, proceeding logically from the general examination of professionalism and technical expertise to the role of technocrats and their relationship to the wider political and administrative system. The analysis seeks to explain the transfer of influence within the nuclear decision-making machinery, away from the experts of the UKAEA and in favour of those at the CEGB, in terms that view such shifts as a result of the political maturation of the Board's engineers/managers.

Chapters 8 and 9 are case studies, designed to extract from the analysis of the preceding sections issues that, when examined in finer detail, illustrate and expand upon the arguments of the thesis, developed from Chapters 3 to 7. To this end the evolution of British Nuclear Fuels Limited (BNFL) provides an illuminating account of the operation of politically aware technocrats in manoeuvring amid the rivalries of Whitehall and the quasi-governmental structures to successfully serve the professional and organisational goal of autonomy in much of their work and policy-making (Chapter 8). The marriage between the political and technocratic areas of government, a continuing theme of this work, is shown by the establishment of BNFL and the role played by professional and organisational ideology in changing wider administrative policies, combined with the importance of winning the collaboration of politicians for the success of these goals. Chapter 9 concentrates upon the result of BNFL's commercial/political success to examine the Windscale Inquiry and its repercussions for technocratic influence in policy-making. The Inquiry represented the dangers inherent from a technical perspective of a loss of expert control over the content of work due to its politicisation. Conversely it was a partial victory for those anxious to increase technocratic accountability. An attempt is made to analyse the political legitimation of a policy through the use of scientific rationality and how that became entangled with judicial rationality. The effects of a division amongst the experts on the proper way to proceed became clear here, a breakdown in the technocrats' cognitive unity facilitating political control. These are valuable lessons which were underlined by the Sizewell Inquiry.

An inherent part of the analysis throughout the book is a concern to expose the realities of organisational politics and to assess the role and influence of the ideology of professionalism in the policy process. But first the attempt at a definition of what is meant by the term 'technical professionals' is the aim of the following chapter, in order that a theoretical base may be established.

FOOTNOTES

1. *New Statesman*, 17/6/83, p.6
2. There is a large and ever-growing literature, but see Dixon (1974) and Conroy (1978) as typical examples.
3. See Blume (1974)
4. See Johnson (1972) p.12

2 The concept of the 'technical professions'

The purpose of this chapter is to attempt a definition of the concept of the "technical professions" and place it within the theory of "professionalism" and its broader social context. There are two general reasons why this needs to be done at this juncture. First, the definition of a theoretical perspective aids analysis per se, because it provides a framework from which an examination of the subject may commence, even if the initial outlines are subsequently reviewed. The necessity of this duty is illustrated by the disparate approaches to the several fields of study contained in this work and outlined in the previous chapter. Second, it is one of the notable aspects of much of the literature on 'professionalism', that it is predominantly concerned with the power of professions such as Law and Medicine or the newer skills that are organised in social work and teaching. It is perhaps possible to categorise these (admittedly somewhat arbitrarily) as the social professions. With few exceptions (such as Prandy, 1965; Gerstyle and Hutton, 1966; Berthoud and Smith, 1981; Laffin, 1986) the specific role of the professional goals of engineers and scientists have been largely ignored in policy studies. There is a need, therefore, to understand what is meant by the term "technical profession".

It is to be hoped that evidence presented in subsequent chapters will provide a step towards remedying this lacuna. But before the presentation of evidence gained in research can be evaluated it is important to construct the theoretical perspective from which to view it in order that a coherence and sense of value may be imposed. Thus, much of this chapter is concerned to explore the basis of "professionalism" and in particular "technical professionalism". In pursuing this task the analysis leans heavily upon the essentially neo-Marxist accounts of Larson and Dunleavy, but it must be emphasised that this does not signal an acceptance of the core elements in such descriptions. Indeed the evidence of succeeding chapters tends not to support the substantive points contained in

Larson's work, which itself is almost a conspiracy theory regarding the class nature of professionalism and its role in the maintenance of the status quo (Larson, 1977, p.156). The utility of such neo-Marxist analyses is as an aid to understanding the role of the ideology of 'professionalism' as a motivating factor in the actions of the professionals and the subsequent possible policy-effects. Obviously this perspective must be linked to the wider social structures.

In particular, the introduction to the literature (in the previous chapter) suggested a relationship between the politics of 'professionalism' and those of 'rationality', a theme returned to later. After the initial exploration of the concept of "technical professionalism" these links are, therefore, examined as a means to expand upon the socio-political implications of the concept in practice. Specifically Dunleavy argues that the professions are a collective attempt 'to upgrade relatively specialised and differentiated (occupational) activities' (1979, p.3). The main factor in influencing the success of this pursuit is the ability of the occupations seeking professional status (or the maintenance of this status) to develop the ideology of 'professionalism' (ibid, pp.3-7). This focuses upon the legitimation of social inequalities and the effective fusion of an educational hierarchy with an occupational one. The ideology 'stresses autonomy and self-regulation' for the professions and this is achieved through state recognition and enforcement of their position in society. These latter points are defended by the professions' claims of the ethical need to protect clients and the need to develop knowledge in a socially neutral and disinterested way (Dunleavy, 1979, pp.3-11). In much of this Dunleavy is following the analysis of Larson (1977).

It will be argued that the entity 'profession' is a dynamic one, with occupations claiming professional status being situated within a continuum between an ideal type of 'pure professionalism' and the pseudo, or quasi, types of techno-bureaucratic occupations (Larson, 1977, pp.178-190). Many of the occupations contained within the ranks of the technical professions are near to the latter end of the continuum, whilst the social professions (at least the powerful Law and Medicine) are nearer to the former end. Larson explains this as due to the evolutionary process whereby the social professions developed in a pre-industrial state and survived in a position of aristocratic patronage, gradually acquiring status through being associated with their patrons. The stratification of the professions into the elite and other ranks dates from this time (ibid). Modern professions are the product of the structures found in liberal democracy, however, and Larson argues that they are a direct result of the collective effort necessary to capture and control the expanded market for their services thrown up by the Industrial Revolution (ibid). The result, she argues, of this 'modern reorganisation of the professional markets' was that the professions 'found credibility on a different and much enlarged monopolistic base', that of a claim to 'the sole control of superior expertise' (ibid). The special conditions that gave rise to law and medicine ceased to exist, yet the ideology created remained and was adopted by the emerging technical groups, including science and engineering. It is argued therefore that not only are there different degrees of professionalism, there are also different types.

A noted feature of the technical professions is their dual allegiance: to corporate employers and to their profession (Dunleavy, 1980, p.111). This is examined in some detail in Chapter 3, but it will be shown that engineers and other technical professions emerged from their training as salaried professionals, not as self-employed autonomous individuals in private practice. The engineers' market for their expertise was, therefore, subordinated from the beginning to the needs of corporate employers, a factor that has important repercussions expanded upon below and in future chapters. It will be argued that the occupational base, that is the organisation of technical experts, is the essential factor in the degree of subordination to which a profession is subjected. This chapter ends, therefore, with a brief discussion of some of the possible implications of technocratic professionalism for policy making.

Professionalism

Analysis seeking to discern a recognisable distinction between the social and technical professions must begin by discussing the essential elements within the professional ideology that make it such a force in many policy-systems. These can be aggregated into two broad concerns: the desire of professionalised occupations to attain and maintain autonomy over the content of their work; a dedication to the goal of achieving and maintaining the self-regulation of the members of the profession , with all that this entails such as a right to control entry to the profession, and codify and enforce a list of 'professional ethics'. It is from these two primary goals that the traits identified by earlier theorists (see Chapter 1) are logically derived. It is one of the tasks of the following substantive chapters in this study, particularly Chapter 8 and its analysis of the evolution of BNFL, to illustrate clearly that the motivation to control work-content has had far-reaching repercussions for policy-making. Further, although this goal is influential within the technical professions, it is clear that for scientists and engineers self-regulation is an ideal which for structural reasons (to be discussed in this and the next chapter) must elude them. This in itself is one of the major areas of contrast with the social professions.

Wilding (1982, p.2) in an echo of Mannheim (1954, pp.77-8) argued that the study of social phenomena could only be understood within the context of the society in which they were found. He called for the study of professionalism to be set within the parameters of an explicit social theory, yet as the work of Saunders has shown (1983), in the complexities of mature capitalist societies it may be necessary to adopt more than one theory to obtain an understanding of social and political behaviour. The introduction to this chapter has explained that the analysis of this study, whilst freely using the tools of at least one (neo-Marxist) approach, nonetheless rejects many of the basic premises contained within that school. The purpose of this assessment of 'professionalism' is limited, therefore, to studying the phenomenon within the confines of the political and economic structures in which it exists in Britain, whilst not attempting to establish an explicit critique of this society per se. In this sense followers of Wilding will probably be disappointed by the limitations of the analysis, yet full cognisance of the debt owed to 'explicit social theory' is attempted. But this is not an excuse to slip into the style of the trait theorists or functionalists, as it is accepted here that professions need to be 'regarded critically, rather than simply accepted' (Wilding, 1982, p.3, quoting Johnson, 1972).

The explanations of Larson (1977) and Wilding (1982) purport to show how professionalism is a product of the structures found within capitalist societies. It is a method whereby the benefits of the collective pursuit of autonomy accrue to members of a profession as individually enjoyed privilege: the material result of the higher status accorded to the professions and their members by society. The pursuit of professional status can be viewed, therefore, as the pursuit of individual privilege. Inherent to much of Johnson's analysis is the belief that one of the reasons why society so readily accedes to the granting of this status is that professionalism is also a method of controlling powerful occupations through the imposition of a code of ethics (1972). It is tempting to accept this argument at face value and Johnson is fairly convincing in his elucidation of what amounts to a pseudo-social contract theory. He fails, however, to take full account of the opposition that many groups within society express towards professional power, a displeasure analysed in some in a later chapter of this book dealing with its manifestation at the Windscale Inquiry. Furthermore, like much of the literature, this explanation fails to recognise the peculiarities of the technocratic occupations. To map these peculiarities it is necessary to expand upon the analysis of Larson (1977).

Social and Technical Professions Contrasted

The process of teasing out the contrasting aspects of the social and technical professions can be traced back to the work of Durkheim (1957), Veblen (in Johnson, 1972, chapter 1) and Merton (1968), although this was not the specific purpose of their work. The most useful of recent aids to the understanding of the nature of professionalism has been an expansion of this earlier work provided by Larson. She has argued that the concept is a dynamic one and should be viewed as though it were a continuum, with a variety of professional and professionalising occupations situated within the parameters of 'routine techno-bureaucratic occupations' and (the ideal type of) 'pure science' (1977, pp.178-190). Larson's development of the concept contrasted slightly with the view of Freidson (1970 and 1973) who argued that the occupational base of a profession is unimportant for deciding whether it can evolve professional characteristics, an approach which tends to be invalidated by the analysis of Larson and the presentation of later chapters in this book.

Because it is situated within a continuum, a social profession may in certain circumstances, display technical characteristics and vice-versa. Furthermore, due to the structural formations of the professions within the corporate sector, technical professions are not able to pass along the line and become part of the social professions: in this sense it may be said that the continuum has fractured . This is a point that is examined in more detail later. It is possible, however, for a profession or parts of it, to pass into the realm of routine occupation and thereby to lose its privileges. It is for this reason that in professionalising occupations like engineering (see next chapter) and nursing, there is a continuing and somewhat acrimonious debate over exactly where professionalism begins. Those in favour of a move towards professional status in these occupations are not aided in their quest by a cognitive and structural unity, a feature which facilitates the claims to status of quasi-professional groups and those engaged in routine work.

Larson's distinction begins with those professions which control their own market for their services and are primarily dealing with their clients on an individual basis. The premier examples of these are Law and Medicine, the major social professions who have been able to subsume their market within their own ideological framework. This is in contrast to those occupations which are unable to control their market and are obliged to operate via the ideological constraints of their patrons in the private and public sectors. Larson calls these 'occupational professions' (1977, p.179) and examples include accountancy, engineering and social work. These groups, she argues, are generated by the expansion of the bureaucratic apparatus of the state (ibid). They fall into two categories, the first of which is:

> "...those occupations that are generated by the concentration of administrative and managerial functions under corporate capitalism. Here the claims of specialised or professional expertise is used for technobureaucratic functions which are unspecific. It borrows from the ideology of professionalism to justify bureaucratic power" (ibid).

Dunleavy provides as an example of this the administrative grades in the British civil service (1979, pp.30-31). Clearly, these occupations are not 'professional' in the sense that the trait theorists or functionalists would understand them; indeed Larson dismisses them as being merely an anomaly, a pseudo-professional structure. They are useful, however, as an example of the power of the concept of 'professionalism' and the realisation amongst non-professionals of the importance of acquiring professional status. This status provides them with the ability to operate within the parameters of professional ideology and thereby reap the privileges accruing, such as control of work content and occupational autonomy.

Larson's second category can be described as containing occupational professions proper. It consists of those occupations that it can be seen are derived directly from the expansion of state functions and attributions, such as teaching and social work, their claims to expertise and professional status are sanctioned by universities and other credentialling organisations. This, she argues, 'provides them with the means of acquiring countervailing power' with regard to the 'bureaucratic organisation within which they are contained (1977, p.179). She continues:

> 'The main instrument of professional advancement, much more than the profession of altruism, is the capacity to claim esoteric skills - that is to create and control a cognitive base' (1977, p.180).

There are two points which can be expanded upon here. First, the creation and control of a cognitive base, a unified body of knowledge, is an attribute without which no occupation can attain autonomy. If this base is fractured (as in engineering and science) or controlled by another group of professionals (as in nursing) then claims for professional status are seriously hampered. Secondly, even should the occupational base possess cognitive unity, if the market for its services is controlled by state or private bureaucratic organisations (over which it has no control) then it is in a position of patronage, and autonomy thereby constrained [1]. Examples of this include teaching, accountancy, social work and some branches of engineering (see Chapter 3 below).

In order to further develop the concept of technical professionalism, a slight reformulation clarifies Larson's categories. Obviously her first category is a mainly managerial one. The members of it have power over each other and over their work because they are managers not professionals, although managers can be, and often are, professionals. In the case of engineers and scientists, people moving into a higher level within their employing organisation become what can be described as 'technocrats' instead of 'technologists'. If sacked, this higher status is naturally lost, although not, of course, any technical expertise, the practice of which is governed by a code of ethics in occupations claiming professional status. Onto Larson's second category, the realisation that it is not only the expansion of state functions and attributes that has led to the growth of occupational professions, but also the expansion, consolidation and generally changing nature of the private corporate sector can be grafted, particularly this sector's reliance upon expertise for innovation and management (Clark, 1985). In some senses this category can even include the old social professions, particularly the requirement of major corporations for in-house legal expertise. In the main, though, the concept of occupational professions in Larson's second category is more usefully identified as those occupations which claim professional status, but which are situated within bureaucratic organisations; these can either be new social professions (such as teaching and social work) or technical professions (such as science, engineering or accountancy). That is, even the social professions here are not usually of the status of full professionalism enjoyed by Law and Medicine (or the clergy), but are professionalising.

At the risk of becoming repetitive, it must, however, be stressed that a constant theme of professional occupations is the search for the goal of autonomy and control over the work content; it recurs throughout subsequent chapters in this study. Chapter 3 examines in detail the position of technocrats in the bureaucratic hierarchies of the state and private sectors, in which science and engineering are located, whilst Chapter 4 concentrates upon the policy-system of nuclear power. It is from the professional structures found here, a position of corporate patronage, that it can be argued the techno-occupations goal of work autonomy is seen to be untenable. This applies not only to the technocrats within the nuclear industry, but to technical professionalism per se. The technical professionals are a product of the structures of the prevailing economic system in Britain. They 'seek the symbols of professional expertise as legitimation for roles that are actually lodged in a bureaucratic hierarchy' (Larson, 1977, p.186). Their emulation of the old established professions, through the adoption of the ideology of professionalism, is misplaced as conditions existing during the formation of those groups and their acquisition of status pre-dated the industrial revolution and the needs of an industrial society which called into existence the technocrats. A point also applicable to the newer social profession. For these historical and functional reasons the structure of the technical professions has led to a lower public recognition of their work and role than that accorded to the social professions (Finniston, 1980): furthermore, they do not interact with clients on an individual basis, but are part of teams which may have a low client orientation (Larson, 1977, Chapters 9, 10, 11). When combined, these structural and operational observations illustrate the fractured nature of the knowledge base, a point studied in more detail in Chapter 3 where evidence is provided to substantiate these observations. This further underlines earlier remarks concerning the nature of technical professionalism being of a different type from Law and Medicine.

All professionals are hired because of their possession of a skill, their privileges accruing from the scarcity of that expertise, and their control over it (Wilding, 1982, pp.19-84). Their sense of professional identity and commitment to an ethical code are the result of a lengthy university-based training. In the case of the technocrats, this professional formation produces them for a role that can be performed only within the structures of the complex hierarchies of the modern industrial state. Expertise is used for the benefit of the employing organisation, not to ameliorate individual ills or wants in the way that the social professions perform their functions [2]. Whatever work autonomy the technocrats may be granted by their employing organisation is as a result of their training socialising them into the mores of the corporate sector; satisfaction of its demands are their reason for existence (Larson, 1977, Chapter 2) : They are not encouraged to view their knowledge base as a cognitive unity: quite the contrary. This situation fosters an attitude of dual allegiance, whereby a technocrat serves both an employer and the goals of a branch of the profession. The career structure of the technocrats further acts to underline this contrast to the social professions. Technocrats are not as dependent as others upon their profession for advancement; promotion is determined by the employer upon the basis of service to the organisation, not professional quality. In some cases, such as the British civil service, this may result in a series of parallel hierarchies being established in order that experts may be separated from administrators or managers. In theory this allows a technically competent but managerially incompetent person to be promoted: in practice, Fulton found that it led to the dominance of the generalist administrators and he called for a more open structure within the civil service (1968). Finniston adopted a slightly different approach (1980), welcoming the promotion of engineers within an engineering capacity. This divergence of opinion has formed the basis of a continuing debate. Certainly it can be seen that in both the public and private sectors, technocrats who seek promotion are often forced to leave a technical post and apply for a techno-bureaucratic position. This is one of the reasons why the two categories, outlined above, are closely linked: these upwardly mobile professionals carry with them into their new managerial position their professional ideology.

Studies by the Ministry of Technology in the 1960s and the Council of Engineering Institutions [3] confirm this movement of the technocrats. Because of their skill and training these mobile professionals retain their sense of professional identity and indeed often contribute to the organisation of their professional institutions. They are at one and the same time both managers and professionals: they form an important communicative link between the leaders of corporate organisations and the technical experts. It is this role that is the subject of a considerable degree of analysis in subsequent chapters, and it is therefore fruitful to examine and summarise the basis of professional power in order to better understand how and why this role is performed. To explain the basis of professional power facilitates an explanation of the way in which that power is wielded to influence policy, and the motivation behind its use.

Professional Power

Wilding's five categories of professional power are commodious enough to cover most inputs into policy-making and implementation that any occupational conspiracy could be capable of. They comprise power in policy-making; power to define needs and problems; power in resource allocation; power over people; and control over the area of work (1982, pp.19-58). These categories have become a standard description of the sources of professional influence and form the basis of his critique which seeks to limit and control professional power. In compiling his model, Wilding acknowledges his debt to the authors of earlier critiques, such as Freidson and Illich (ibid). Simply to reiterate his arguments would be of limited value for this study: it is perhaps more apposite, therefore, to restrict a review of their applicability to the power of the technocrats -a process that begins to illustrate the links with organisational, or bureaucratic, politics as discussed later in this chapter and referred to in Chapter 1 and which informs the analysis of the subsequent chapters.

In his case studies of the health-care professions, Wilding illustrates how the medical profession has powerfully affected the development of the provision of health services in order to facilitate the realisation of professional goals, rather than to advance the public interest (ibid). Services have been organised to promote the division of occupational skills instead of adopting a holistic approach which, it could be argued, is more suited to meeting client needs (ibid, pp.23-27). In the case of medicine, a prestigious and united elite have had a direct and substantial impact upon policy-making and this power has been used to achieve a professional advantage. In the performance of the technical professions, however, their position within bureaucratic structures impedes this robust interpretation of professional interest due to their subservient role and the dual allegiance of their members. It is organisation theory that begins to provide a clue as to their power in policy-making. The work of Simon (1957) and Downs (1967) suggests that because of the bureaucratic nature of the technocrats, their strength in shaping policy could come from their key position in controlling information flows. The style of its presentation or the withholding of information are an essential prerequisite to the choices decided upon by policy-makers. There is a further source of power accruing to the technocrats here, and one that is referred to in the general propositions of Chapter 1: because the technical professions are spread across a welter of bureaucracies, they are able to influence policy at several different levels and positions. That is, the professional organisation functions as a conduit of across-institutional barriers, an activity Dunleavy describes as 'ideological corporatism' (1981, pp.3-15). The utility of this concept is examined in the chapters which follow, although it must be remembered that such behaviour is also an example of professionalism per se and is most pronounced in Law and Medicine.

It is the power to define a client's needs and problems which lies at the heart of professional influence. In the realm of the social professions Wilding builds upon the work of Illich to show the 'cognitive imperialism' this leads to, ultimately 'disabling' the individual by making him wholly dependent upon the professionally qualified (Wilding,ibid, pp.29-35). The contrast with the technocrats is evident in that their clients are organisations not individuals. Yet professional attachment to rationality through the ideology of professionalism and its emphasis upon empiricism and quantitative legitimation has led to policy-makers to define complex social and political questions in technical terms. The nuclear power industry is an example of the technocrats' attempts to so define policy-option and the following chapters tend to illustrate Ellul and Borguslaw's argument that the transformation of a political issue into a technical problem is not a neutral process (1981). An example examined in some detail below is the Windscale Inquiry and here it can be seen that in their attempts to set the problems facing reprocessing within technical parameters it behoved the technocrats to 'discover' through the scientific method the 'one best' solution.

Of Wilding's other categories, power in resource allocation and power over people are inextricably linked with organisational and management structure when viewed in the context of technocratic power. These sources of power are perhaps less developed within the technical professions and in any case are overlapped to the point of obscurity by the function of problem and need definition. It is in control over their area of work, a logical progression from the preceding categories, that professions naturally strive the hardest, the attainment of this privilege being one of the fundamental goals of professionalism. As the acme of professionalism, it can be said to have been truly reached only by Law and Medicine, both of which exercise a large degree of control over other occupations in addition to their own field of work. It is an inherent aspect of the preceding argument that the technical professions do not possess such autonomy, although that does not prevent them from seeking it. Their structure, despite acting as a barrier to this power, can paradoxically assist them in influencing work content via their influence within their employing institution if that organisation is itself a professionalised body such as the CEGB and UKAEA (see Chapters 4 to 7 for a further explanation). If the employer has a technical role, such as the provision of electricity on a large scale, then the influence of the technocrats is enhanced as theirs' is a major voice in the policy decisions. As noted, this aspect of professional power is analysed in full in the following chapters, where it is argued that only through the domination of their employing organisation can the role and influence of the technical experts in shaping policy be realised, and that presupposes their employer is itself a powerful actor in the policy process. Before this can be returned to, however, a brief reiteration and expansion upon the links with the wider literature that the above discussion has revealed will serve to aid an understanding of the wider socio-political context within which the technocrats are situated. A task that prepares the way for a fuller exploration of their role within that content in subsequent chapters.

Rational Administration with a Corporate Bias

In referring to the literature on 'empiricism', 'rationality' and 'science', one is constantly reminded of Peter Self's description of it as 'nonsense on stilts' and his demolition of its manifestation in the guise of 'cost benefit analysis' (1975). A treatise which obviously did not convince Layfield, who called for Cost Benefit Analysis to be included in power station siting analysis (Sizewell Report, 1987, Chapter 2). The pull towards rationalist thought within the professional ideology remains as powerful, however, as it does in other spheres such as policy analysis (Carley, 1980). Smith shows that the acceleration of technical influence in decision-making and the concomitant reliance upon 'scientism', dates largely from the corporate structures established during World War Two and the immediate post-war era (1979). The influx of expertise into Whitehall irrevocably altered the processes of administration. Whilst the increasing demands of the evolving Welfare State led to a coalescing of society into a complex series of corporate hierarchies and a network of promises, commitments, obligations and expectations that ebbed and flowed between the state and public and private sectors. These locked the politicians into a constricting range of options and were, Smith argued, a limitation on the democratic process of the party system, leading to what Middlemas has called not a corporate society, so much as a corporate bias in society (1979). In this the rise of the professional ideology and its a-political emphasis upon empiricism has been important.

Professionalism contains a strong element of positivism which views knowledge, or science, as progressing through a logical process of hypothetico-deduction (Blume, 1974). Popper viewed science as being creative and critical, marking out the world as a place of observable phenomena subject to general laws. A knowledge of these laws would, therefore, aid control of the phenomena and it is the task of science to obtain that knowledge (Blume, 1974, pp.32-57; Fay, 1975, Chapters 2, 3 and 4). Scientific knowledge exists as 'theory' unless or until proved faulty when it is then reformulated, ensuring each theory is superior to (or more correct than) its predecessor (Blume, ibid p.32). Kuhn developed this to argue that scientists should be viewed as being a part of a community (1970). There are numerous communities, each operating within its own paradigm: a scientist produces for the community to which he belongs and it is this which validates his findings (ibid). Blume interprets the nature of change within this paradigmatic universe to occur only 'when problems appear that resist solution in the light of the currently employed paradigm' thus casting doubt upon its value and necessitating the construction of new hypotheses (1974, pp.35-48).

This community-validation function is obviously at work within the professional ideology and it is this which legitimises the advice tendered to policy-makers. A united front by the experts provides little room in which a technically ignorant politician, civil servant or managing director, can manoeuvre in order to resist technocratic pressure in favour of a particular option. This provides a theoretical grounding for one of the lessons of the following chapters, that is the resulting loss of professional power should that unity be fractured. The function of empiricism as a basis of professional advice is important because as political and social decisions have increasingly come to be defined by experts, particularly when they exhibit a cognitive unity, so many of the policies under which society is administered have similarly come to be based upon an empiricist foundation. In contrast, traditional legitimisers such as moral and religious yardsticks have declined in importance as a basis for decision-making. Theories abound which seek to take this process further and eliminate what is seen as unsound subjectivity in the policy-making process (Carley, 1980; Rein, 1976). From this process is drawn the concept of 'scientism'.

Wynne identifies scientism, which apart from semantic hair-splitting is empiricism by another name, as the legitimator for many aspects of political decision-making in the nuclear policy-system (1982, p.11). He argues that the use of scientism in this way has secularised and diminished its authority, undermining its claim to be rooted in objective quantifiable data found in observations of an empirical world (ibid, p.96). In its study of the Windscale Inquiry a later chapter illustrates how as scepticism of the objectivity of the technocrats has grown as their unity has diminished, the even more extreme rhetoric of precise rationality and 'elaborate formalisms of the judiciary has filled this gap' (Wynne, 1982, p.96). It has done this by 'projecting renovated images of expert discovery' and cloaking this in the 'superior (legal) expert resolution of Inferior (scientific) expert conflict' (ibid). Wynne, therefore, identifies the essential rationalist foundation of the professional ideology as being employed to settle disputes between sparring professionals, a process clearly linked to the political and commercial demands of the corporate economy.

Implicit in much of Wynne's analysis is a sense of irony that judicial rationality insists upon absolute certainty from science as the Law exhibits a 'growing use of scientific expertise' and cultivates its own authority by 'reference to empiricist models of natural science' (ibid, p.129). The irony is compounded by the realisation that 'scientific knowledge is a fragile, shifting network of interpretive and theoretical activity', but that this process of negotiation goes unrecognised by the legal mind (ibid). Even what can be considered to be an empirical observation may be the "subjective interpretation of experts" (ibid, p.131). It can be argued, therefore, that the claims of the professionally trained to posses objective, empirically verifiable solutions to objectively defined problems are somewhat shallow. The analysis of Chapters 4 to 7 in particular, but also the case studies, attempt to illustrate this and a dawning realisation by the politicians that the notion of professional objectivity was flawed in the respects outlined above. Technical advice, therefore, reflected the bias of the organisation's experts who were responsible for tendering it within the nuclear power policy community.

The repercussions of this for professional power are apparent, as Larson as attempted to argue, the professions evolved through uniting their members around a cognitive unity and science proved to be the most collectively acceptable agent (1977, chapters 1, 2 and 3). This allowed them, in the case of Law and Medicine, to achieve both structural unity and maintain their status with the public. Science, being outside of the understanding of most people, was a beneficial and esoteric force to which society deferred. As the new priests interpreting the scientific method of empiricism, this deference was focused upon the professionals. To challenge the basis of this deference is to challenge the basis of professional power and threaten the interests of the organisations on whose behalf that power is exercised. The challenge to professional power in policy-making has come from across the political spectrum and has reflected a desire by some to escape from the confines of technocratic decision-making and achieve a wider or more political approach to this process (Smith, 1979; Wynne, 1982; Habermas, 1979; Middlemas, 1979). An attempt to widen the horizons of legitimate political action is a challenge to professional power (Ellul and Borguslaw, 1981). Marcuse has even argued that 'the whole structure of scientific thought is now seen as logically with - and indeed based upon - the idea of domination' (in Blume, 1974, p.53). Whilst there is not the evidence to suggest the 'whole structure' is so connected, certainly the above comments regarding the links between the rise of professionalism and the expansion of the corporate bias in society appear to have some foundation.

The massive literature concerned with corporatism is so varied that there are grounds for considering the concept no longer possesses utility as a precise term in political science, despite Schmitter's efforts at definition (1974; 1977). It is useful, however, when viewed from the perspective of Dunleavy, when he applies many of the arguments outlined above to the policy process under the label 'ideological corporatism' (1981, pp.3-15). In this sense he uses the term to show the importance of the professional ideology and structure to subvert the inter-and intra-organisational boundaries of the public and private sectors, an approach analysed episodically throughout this study. A second important use of the term is found in the work of authors such as Smith (1979) and Middlemas (1979) in their explanations of the developments of the British economy and society. As noted above, their (slightly differing) interpretations of the reorganisation of the socio-economic structures into hierarchically ordered and consensus-orientated organisations operating under conditions of reduced competition, accepted a view of corporate bias (whilst not agreeing with it) under which professionalism thrived. Indeed it can be argued that certainly within the Electricity Supply Industry, its nationalisation established the conditions necessary for the expansion of technocratic influence.

These broader considerations regarding the position of professionalism are essential to an understanding of its role within the policy process of nuclear energy. This is not to

"...bind together blocks of organisational analysis in the fashion of a child playing with Lego..."

which Gray and Jenkins warn against (1985, p.29), but to display an awareness of the fiendish complexities inherent to any such study. Within the field of administration itself, the perspective must be dynamic in order to be 'concerned with the process of change and its origins' (Pitt and Smith, 1981, p.10) and the very real but increasingly subtle nuances that distinguish the public sector from the private. There is also an awareness that:

"...the world of a higher civil servant is characterised by both inter- and intra-organisational differences. Moreover, it is not a world of peaceful hierarchies where juniors simply carry out the instructions of their superiors" (Gray and Jenkins, ibid. pp.33-34).

These points are relevant to organisations per se, and it will be shown in the following chapters that the role of the technocrats in the internecine manoeuvrings of Whitehall, the AEA and the private sector, has been one of considerable, though varying, import.

Gray and Jenkins are particularly instructive on these issues when they argue (following Pfeffer, 1981) that as most complex organisations change 'they re-define existing or take on new tasks', a process evident in the evolution of BNFL. This structural differentiation is often associated with identifiable 'sets of values and interests' (Gray and Jenkins, ibid, p.34). Although referring to Whitehall, it is clear that this process can apply to organisations per se and that in many cases these identifiable sets of values and interests can be closely linked to the ideology of professionalism, particularly if Dunleavy's concept of 'ideological corporatism' is sound (op.cit.). Gray and Jenkins have seen professionalism within the public sector in terms of being sophisticated codes of accountability (ibid, p.145). This would certainly fit with the professionals' own claims to possess a code of ethics, but the authors' points regarding the importance of groups of policy actors identifying with each other and seeking autonomy within the policy community, are also of value (ibid. pp.30-44) paralleling as they do many of the traits normally applied to professionalism. This suggests, therefore that goals such as 'autonomy' are inherent to organisations and that professionalism should be viewed as one of several ways of attaining these goals, with the ideology of professionalism being far from unique in its ultimate ambitions. This would certainly go some way towards explaining the convergence of organisational and professional goals inherent to the bureaucratic position of technocrats. Technical professionalism should perhaps be viewed, then, as a cognitive organisation, situated within and across the hierarchies of the public and private sectors. As such it is both subservient to and informs these structures.

Policy Implications

The last perspective is one that is generally adopted for the rest of this study. It has several implications that expand slightly upon the general propositions outlined in Chapter 1. The first point is that when combined with the organisational goals exhibited by actors in the policy process, professionalism can facilitate an increase in the autonomy of agencies. Dunleavy argues that this propensity towards autonomy via professionalism is particularly acute in the case of single-issue organisations like the UKAEA (1979, pp.9-11). In these agencies the dominance of a single group of professionals may lead to an overemphasis on the need for activities which reinforce the professional role. Prandy argued, for example, that the scientists employed by research organisations tended to over-emphasise the importance of pure research as this increased their occupational control (1969), an activity that certain parts of the UKAEA (Chapters 7 and 8) were certainly prone to. This parochialism tends to be endemic to group activity and is one of the reasons given by generalist civil servants (see below, Chapters 4 to 9) for the need for them to retain their somewhat discredited role as unbiased analysts able to take the wider view.

A second expansion upon the opening propositions is that it can be argued that the growth of professionalism serves as a tool used by the technocrats to consolidate their dominance over some policy area. Again, Dunleavy foresaw the possibilities of this (1979, pp.19-35), but he tended to limit them to the activities of management groups to professionalise themselves. It can be argued that the ideology of professionalism has been carried by the professionally qualified into the techno-bureaucratic occupations where, combined with organisational perspectives, it has served as the foundation of technocratic arguments over policy options. Specifically a belief in the scientific method to discover superior solutions to problems defined in technical terms, but which were previously thought to be of a political or social nature. This ability is dependent upon the cognitive unity and indeed the structural unity of the professionals. Where this has been undermined an important implication has been the reassertion of political control, an activity misunderstood and resented by the professionals as a threat to their autonomy. Examples of this reassertion include the later reactor decisions (Chapters 6 and 7). But it must be remembered that other factors are also involved in these procedures, in particular the efficaciousness of professional options as solutions to problems, and the broader canvas of bureaucratic politics upon which the whole system is pinned: these points are examined in some detail in the following chapters.

Finally, the discussion of this chapter has explicitly enunciated a view that technical professions exercise an influence upon policy via their employing organisations (or patrons). This is due to their position within the bureaucratic hierarchies of the socio-economic system. It follows from this that the more influential technocrat-controlled agencies are in the policy-making process, then the more that the options chosen to be pursued will be those favoured by the technocrats themselves. This is not to suggest that technocrats explicitly seek to control organisations in order to advance their own options: quite the contrary. Their position in a state of corporate patronage ensures that it is the requirements of organisational goals (however or by whomsoever these were set) that determine the exploitation of professional skills. It is only if the original goals become redundant that a reformulation has to occur in order to maintain the group, not the profession.

In order to examine these points in more detail, the theoretical arguments may be expanded with an analysis of the practical realisation of technical professionalism which aids an understanding of the content within which nuclear experts work. To this end, the next chapter reviews the structure of the engineering and scientific occupations, assessing their positions upon the professionalism continuum. This prepares the ground for Chapter 4 which first plots the position of the technocrats within the nuclear power policy system and then begins the analysis of their role and influence in the choice of policy options. The discussion of this chapter has been an attempt to highlight the relationship of this study to the broader framework and chart the course to be followed.

FOOTNOTES

1   For an introductory discussion on the notion of patronage see Dunleavy (1979).

2. See Fulton (1968); Self (1972); Kellner and Crowther-Hunt (1980) for a view of the result of this in the civil service and public sector; Finniston (1980; Prandy (1965); Berthoud and Smith (1980), for explanations of its impact on the private sector.

3. Mintech surveys were 1966 and 1968. (CET surveys 1971, 1972, 1975, 1977, 1981).

3 Science and engineering

It is the intention of this chapter to use the previous general theoretical discussions in order to focus upon the occupational structures of scientists and engineers, and then to assess the extent of the applicability of the concept of "technical professionalism" therein. The following sections, therefore, will examine the problems incurred in defining "science" and "engineering", before attempting to do so, and then proceed to explore the relationship of the technocrats to their workplace and how this may affect the policy process. The justification for adopting this approach can be seen in the desirability of understanding the context in which the technocrats are situated in order that the motivation behind those activities affecting policy are clarified. That is, technical professionalism as discussed in Chapter 2 can be seen as a broad sector on the occupational continuum, but this book is concerned specifically with those technocrats who are explicitly involved with influencing nuclear policy.

In this Chapter, therefore, there is an examination of the influence of "professionalism" upon scientists and engineers, as they are the dominant technical professions in the nuclear industry. In order to aid analysis of their influence they are first described in terms of a manageable analytical concept, a process that involves an assessment of the applicability of the previous chapter's arguments. The analysis is then ready to proceed towards exploring the relationship of these technocrats with their environment, how they are shaped by it and in turn attempt to change it; a vital step on the way to the substantive part of the thesis which examines their influence on nuclear decisions. Only when the subject under investigation is so clarified is it possible to proceed.

Definition is the first and most persistent problem confronting the researcher engaged upon a study of scientists and engineers. There is no universally accepted definition of either occupation. Indeed, because there is no definition, there are no reliable descriptions of exactly how the work content of these people should be construed, or even how many scientists and engineers there actually are (Berthoud and Smith, 1980). The problem is compounded when it is realised that in private industry the distinction between the two occupations is at best blurred and frequently cannot be made at all. Many firms simply do not classify their employees as scientists or engineers. It is argued that in the public sector and universities there is a distinction made for example by separate faculties, and it is here that the occupations are at pains to emphasise their differences. The analysis of the previous chapter suggests that these problems are as a result of the lack of cognitive unity found within these occupations and (it is worth repeating) their position as part of the hierarchies of their employing organisations, two points expanded upon below. Such a situation would appear unthinkable within the present organisation of Law and Medicine, but definitional difficulties are part and parcel of Science and Engineering, two severely divided occupations.

It will be argued below that it is a misnomer to refer to the occupation of 'science' or 'engineering'. 'Science' is not an occupation as such, but a generic term describing a series of occupations, some of which are professionally organised. The organisation itself is a diaspora, scientists have a common denominator in their scientifically-structured pursuit of knowledge, but not a common organisational structure. Likewise, 'engineering', it is argued, is a blanket term to describe what can most charitably be called a 'confederal' structure of organisations, the leading members of which are professionalising, but not professional. The major disciplines (civil, mechanical and electrical engineering) have developed a greater degree of professional status than some of the smaller groups and it is these (along with the chemical engineers) who have the most relevance to the study of nuclear power. Certainly it is to be expected (and will be argued) that the professional elite, which (as noted) is usually simultaneously the managerial elite, acts as an important medium of communication, transmitting and interpreting technical goals to the political policy-makers and vice-versa. The following sections will begin the explanation of this, preparing the way for a fuller analysis in future chapters.

Problems of definition
The technocrats' own definitions

Both engineers and scientists tend to use the descriptive and normative terms of trait theory to define the professional aspects of their occupations [1]. They accept the enunciation of professional ideology employed by the social professions and seek to apply this to themselves in order to achieve professional status and privilege. This has caused some considerable dissent within the ranks of the engineers as institutions seeking to raise the status of engineering have attempted to restrict the description of 'professional' to the more highly qualified sectors. A move that has led to some resentment amongst those 'defined out' and confusion about the perennially blurred boundaries between 'engineer' and 'technician'. In its study of the engineering occupation, the Finniston Inquiry tried to resolve the question of where professional status began by adopting the definitions of the trait theorists Cogan (1953) and Millerson (1964) and interpreting the boundaries fairly liberally. The committee argued that the great majority of engineers were professionals because they fulfilled the three main characteristics of professionalism identified by the literature; that is:

(a) they are required to be expert in a particular field of activity, for which an advanced and extended formation is necessary, and practice in which requires a high level of theoretical foundation;
(b) they have custody of a clearly definable and valuable body of knowledge and understanding;
(c) they accept responsibility and accountability for the decisions they make against recognised values and standards of conduct (Finniston, 1980, p.125)

It will be argued below that at least the last of these does not apply in practice and the first two are negated by the positioning and structure of the engineers and scientists within the corporate sector.

Finniston anticipated these criticisms by recognising that:

"...the great majority of engineers are employees of companies and other organisations; the nature of engineers' work is not well understood by the lay public with whom they are seldom in direct contact; and the range of activities covered by engineers is greater than for most other professions" (ibid).

Whilst recognising the flaws in the definition Finniston did not, however, attempt to tackle them in a theoretical manner, preferring instead to concentrate upon recommending a series of changes to the training of engineers and the structure of the occupation that it was hoped would result in a higher public status. Much of the low public esteem in which engineering was held, argued the report, was due to a lack of public awareness of the true nature of the engineer's work and, therefore, the remedy lay not in drastic changes to the formation of engineers but in educating the public to acknowledge the status of engineering to a similar extent to that of 'doctors and lawyers' (ibid).

The Finniston approach was a continuation of a long tradition of trait definitions by the technocrats themselves. From the early 1960s the Council of Engineering Institutions (CEI) had adopted this interpretation of the professional ideology, showing a particular concern for the espousal of a code of ethics in imitation of the social professions. Most of these attempts at definition fell back upon rather fulsome (perhaps even wishful) descriptions of an engineer's role:

> "An engineer is a man who applies resources of men, money and materials to mould the physical environment and produce the machinery and goods required by society...he will always be a planner, and will frequently be called upon to exercise managerial and perhaps financial skills...the professional engineer is the synthesiser par excellence, the modern creative man" (CEI, 1966, p.2).

Undoubtedly there is a large degree of propaganda in such hyperbole, but it reflected a genuine grievance felt by the engineers regarding a lack of recognition of their skills. Many senior officers in the Professional Engineering Institutions have tended to eschew further detailed attempts at definition opting for simple descriptions instead:

> "It is very hard to define a professional electrical engineer, as it is to define a professional medical man. You do not define them really, you describe them"

argued an officer from the Institution of Electrical Engineers (IEE).

Attempts at defining the professional identity and role of scientists have tended to be more complex, reflecting the nature of science itself and confusing the industrial applications of applied science and the 'pure' role of academic research in universities and other organisations. The ideology of 'scientism' and its links with the ideology of 'professionalism' (see Chapter 2 above) have added to this confusion, aiding the search for professional status at one end of the scientists' spectrum, but of little utility within the hierarchies of private industry where, as noted above and discussed below, the distinction between scientists and engineers is not a valid one. One leading engineer opted for a simple definition, arguing that it is:

> "...observational, in the sense of directly looking at nature or setting up experiments to look at what nature does and to draw conclusions from that. To construct generalised theories that will allow you to predict future phenomena which can then be sought and if found either support the theory or not".

The links with the empiricism of the positivists are fairly explicit in this definition. Added to this is the concept of it as a community or series of communities, in the manner of Kuhn's interpretations, and reflected in the work of Dixon. He calls for science to be viewed as a 'brotherhood' remarking favourably that the 'ethics and values' of science are one of its 'crucial aspects' as a social activity (1973, p.46-59). The concern with empiricism remains, but this is linked to 'a satisfying social role and a defined career structure' (ibid, p.13). Quite how this is achieved is not explained, and indeed cannot be explained as there is no definitive professional structure within science, any more than there is within engineering. This was recognised by the Council for Science and Technology Institutes (CSTI), which stated in bland tones that in its view:

> "...a professional scientist is any person who, on the 1st of April, 1968, was a Fellow; Associate Fellow; Associate or Member; Licenciate; or a Graduate Member of one of the corporate members of the CSTI" (CSTI and Mintech, 1968, p.1)

These somewhat vague (and sanguine) descriptions are of only limited value in providing an aid to understanding the nature of engineering and science. It is more helpful for this thesis to scrutinise the technocrat's own definitions with more rigour from the context of the analysis of Chapter 2. This is in order to understand the broader construction of these occupations, before moving on to study their position within the nuclear policy framework in the next two chapters. Such an understanding providing a perspective from which to approach the policy process.

Technocrats: a re-definition

Engineers

The 'confederal', even splintered, nature of engineering can be glimpsed when it is realised that there are about eighty various engineering institutions in existence nationally. Watson traces the history of this tangled structure from the formation of the Institution of Civil Engineers (ICE) and Institution of Mechanical Engineers (I.Mech.E) in 1818 and 1847 respectively, until the process of easy fission evolved into the present complexity (1975, Chapters 3 and 4). He argues that this is perhaps the occupation's major barrier to full professional status.

The professional engineering institutions (PEIs) perform a dual function in their capacity of representing the interests of their members. The role of 'learned societies' is their first duty, and a member of the Finniston team argued that:

> "...The most important role played by the institutions is that they have great educational value. Although people tend to underrate the need for education, the world is changing extraordinarily rapidly. They are learned societies and I don't see why they should attempt to be other than that".

In executing this role, seminars, major publications, newsletters and ad hoc committees of inquiry are the media by which the institutions seek to educate the membership. This network of communication cuts across all organisational barriers and is closely connected to the second function the institutions perform; legitimation of the attainment of professional status for the discipline and members they represent. The sixteen corporate members of the now defunct CEI achieved this by awarding the status of Chartered Engineer (C.Eng.) upon novices who satisfied the stated requirements. The PEIs have retained this through the demise of the CEI and its replacement by the Engineering Council, following the recommendations of the Finniston Report (1980). All the institutions award titles such as 'Member', and 'Fellow', which it is claimed bestows a professional recognition upon the individuals concerned.

The 'professional legitimation' role is one that has led to criticism being levelled at the PEIs [2]. In contrast to the established social professions, engineering's fragmented base illustrates the lack of the cognitive unity identified in Chapter 2 as an essential prerequisite for the mounting of a campaign to secure professional status. The plethora of PEIs, each seeking to further its own limited professional goals, has severely hampered the search for a united professional identity. Watson argued that each fission represented 'groupings of would-be professional engineers' who have 'tried to usurp the status of the established elite, throwing it into question' (1975, p.91). This absence of professional agreement with regard to the boundaries of the professional community has, he argued, led to a refusal by society at large to confer professional status upon groups 'the legitimacy of the claims of which have been denied by those who should be most aware of their validity' (ibid). Members of the Finniston Inquiry had some sympathy with this view, but pointed out that the spectrum of the occupation is so wide that it is 'administratively correct that the various classes should be looked after by specialist institutions' [3]. The amelioration of this disunity should, they argued, be through the establishment of an 'Engineering Authority' which would pursue the common interest in the manner of the medical profession's General Medical Council [4]. Such a notion has not found favour either with the engineers or with the Government: indeed, when the account of the engineering structure in the nuclear industry is studied (Chapters 4 to 8) it is difficult to envisage how the Authority could ever be practicable, such is the lack of a genuine community of interest. It can also be argued that Finniston's own analysis does not lead logically on to the conclusions he reached with regard to the need for an Authority [5].

It would be fair to assert that the proposals of the Finniston Report and the goal of greater professional unity and status elicit a mixed reaction from within engineering, hardly surprising considering the complex nature of the occupation. There does appear to be, however, a general desire from the engineers for a greater recognition of their claims for professional status and a realisation that this is best served by a collective approach to the situation. This hankering is a well-established facet of engineering, and indeed of all professionalising occupations. It was the desire for a greater unity which prompted the establishment first of the Engineers' Guild and then its replacement in 1962 with the CEI. Awarded its Royal Charter in 1965, the CEI was a federal body of which the sixteen largest PEIs were corporate members and to which a further eleven smaller PEIs were affiliated. In claiming to represent over three hundred thousand individual members (a claim made slightly awry through the regular practice of double membership) the CEI stated its aims as being:

> "...to promote and co-ordinate in the public interest the development of the science, art and practice of engineering" (CEI, MINTECH, 1966, p.1)

To this end it sought to 'establish, uphold and advance the standards of qualification, competence and conduct of professional engineers' (ibid). The fostering of relations with government, the universities and other public bodies was a further goal aimed at raising the level of official awareness and (hopefully) the status of engineering,. The CEI set up the Engineers' Registration Board in order to register all Chartered Engineers, and generally sought the elevation of the occupation through this emulation of the social professions' methods of operation. It was (and remains) an example of the utilisation of the professional ideology to accrue a level of privilege that is not commensurate with the organisation of the occupation.

With hindsight, the practice seemed doomed to be the failure that it was. The old rivalries remained and the CEI became something of a front for the 'big three' (ICE, IME, IEE) on which it depended for the bulk of its resources and who continued to dwarf the remainder of the PEIs. The IME's 'Dawson Report' (1978) and the IEE's 'Merriman Report' (1978) were severely critical of the progress towards greater unity and status. These were joined by a growing clamour from within the membership, which was increasingly seeking trade union membership as an alternative method to pursue individual interest [6]. The Electrical Engineers' Association (EPEA) added its voice to the fray, calling (with others) for the Government to intervene in the organisation of engineers. The EPEA did not believe that the CEI and the PEIs were capable of sufficiently improving the status of engineers and presumably, therefore, their conditions of work .

The result of the pressure brought to bear upon it, combined with its own unfavourable impressions, led the Government to establish the Committee of Inquiry under the Chairmanship of Sir Monty Finniston. As noted above, his main recommendation was the establishment of an 'Engineering Authority' to be what he called 'an engine for change' (1980). The Government and in particular the Industry Minister, Sir Keith Joseph, were committed to a reduction in both the number of quasi-governmental agencies and the corporate bias inherent to the economy (see the previous chapter above). The desire of the Inquiry to have the Authority made a statutory body did not, therefore, find favour and the result was a dilution in the form of the 'Engineering Council', established in 1982. The Engineering Council (EC) is a non-statutory body, although like the CEI it has a Royal Charter. It is too early to comment upon the EC's attempts at reorganisation, but there is some substance to the belief that it will suffer from the same inability to impose its decisions as did the CEI; thus an opportunity to acquire greater professional status was lost when the Government refused to bestow statutory powers upon the Council .

The Engineers' Workplace

The first point to make with regard to figures that purport to show the deployment of engineers in Britain is that they are probably inaccurate. This is because most of the surveys so far conducted have used the membership lists of the PEIs, but these do not represent all of Britain's engineers as professional status is not dependent upon the title C.Eng. and the UK does not practise the licensing of engineers. Furthermore, PEI and CEI surveys have not met with the full cooperation of their members, returns being disappointingly low. Finally, those surveys that have been conducted have not compared like with like: for example only the later CEI efforts were concerned with technicians, whereas Berthoud and Smith's figures obtained for the Finniston Inquiry contained technician engineers, despite the protestations of the PEIs that these were not 'proper' professionals (Berthoud and Smith, 1981). This dearth of demographic information is a problem that the Engineering Council is acutely aware of, estimating that in Britain there are at least a quarter of a million engineers and seeking (as yet unsuccessfully) new methods of quantifying them accurately. The best that can presently be hoped for is an outline of the distribution of engineering expertise as is provided by the bi-annual salary survey, the 1985 edition providing greater detail than previous ones, although even this only had a response rate of 43.7 per cent (Engineering Council, 1985). Figures produced by the Engineering Industry Training Board (EITB) in 1986 purport to show nearly 86 thousand professional engineers, scientists, and technologists in Britain; nearly 67 thousand of these being professional engineers (EITB, 1986 (a), Table 6).

As expected, the largest percentage of engineers are located within industry, the approximate ratios remaining fairly stable over the period 1962-1980, within the various sectors. Between fifty and fifty-five per cent of engineers are employed in the private sector, between two and four per cent are self-employed here,(although the 1985 EC survey suggested nearly 7 per cent) and the rest are found within the local authorities (nine to fifteen per cent), public industries and utilities (about fifteen per cent) and universities (about three per cent) [7]. With the move towards greater privateisation these ratios have begun to alter slightly and the EITB's 1985 figures show that about 60 per cent of professional engineers,scientists and technologists, "are employed in just three sectors: electronics, aerospace, and office and data processing equipment." (EITB, 1986 (b), Summary). Clearly, with so few engineers self-employed or situated within the universities, the standard position is for an individual to be a part of a team located within the hierarchies of the public or private corporations or a local authority: a position which tends to support the analysis of Chapter 2 with regard to the nature of technical professionalism and its contrasts with the established social professions.

The Finniston survey illustrates that as engineers progress in their careers they tend to leave the applied side of their expertise and gravitate towards a managerial function. Berthoud and Smith found that by the age of fifty nearly half of all engineers were in a managerial role (1981, p.2). Of these, at least a third were no longer actively using their engineering expertise in the execution of their duties (ibid p.48). The CEI figures tended to suggest an even greater shift towards management (sixty per cent in 1973), results supported by Gerstyle and Hutton (1973, p.17; 1966). Finniston saw this haemorrhaging to management as a disturbing facet, arguing that many young engineers currently employed in primarily technical functions foresaw themselves progressing in their careers in posts in which they would not directly be applying their engineering knowledge and that career patterns confirmed a small but steady shift from engineering practice, mostly in management and teaching positions' (1980, Chapter 3).

The engineers themselves claimed that they were 'forced to quit technical engineering jobs' and enter general management or other jobs in order to advance in their careers and improve their standard of living (ibid). Finniston asserted a need to provide an adequate career structure able to retain this talent within engineering itself (ibid, list of recommendations). This, of course, is in contrast to the recommendations Fulton made for the civil service (1968) where he called for the active migration of technical expertise into general administrative positions. In taking a strict position on the interpretation of 'professional engineer', the EITB goes so far as to argue that managers or technical directors who possess standard chartered engineer qualifications, but are nonetheless not engaged on technical work, should not be included in the definition (or survey of) professional engineers (1986, p.1). Such a purist perspective is not generally accepted, certainly within many of the PEIs or large technical organisations like the CEGB.

Whatever the costs of benefits accruing to either side of this debate, the salient consideration for this study is that this managerial shift reflects a point made in the previous chapter: namely that technologists imbued with a desire for professional status and recognition move into technocratic positions carrying with them the goals of the ideology of 'professionalism'. They become 'engineer-managers' and thus occupy a dual function in that they not only represent the higher management of their employing organisation; they are also part of their occupational/professional elite. Officers within the PEIs have argued that their 'professional leaders' are drawn from throughout the upper echelons of the public and private hierarchies [8], further corroborating earlier assertions (above, last chapter) regarding the role of professional structures as a cognitive organisation cutting across formal institutional barriers in the public and private sectors. These leaders form the conduit through which technical advice and technocratic options are channeled into the deliberations of the policy-making elite, and vice versa. These individuals are collectively the medium of technical influence and are, therefore, a focus of the following chapters concerned with nuclear policy-making being vitally important to that process.

Considering their importance, surprisingly little is known about the process of promoting these individuals (an aspect also found in the following chapter with regard to the nuclear experts) or indeed of their social and educational background. Because of the almost complete lack of reliable information on these matters, any but the most general observations would be fool-hardy. The few crumbs of data that are available suggest that within the nuclear policy system those technocrats who attained a position of influence tended to be better educated and from a higher social background than engineers generally [9]. But such observations are almost cliches and are in any case too vague to be of much importance. This is certainly an area of research that would benefit from a comprehensive study.

Similarly, demographic information on the great mass of engineers is also poor. The results from the few sporadic and limited surveys suggest an increasing tendency towards a middle class all-graduate background. The EITB's occupational profiles, begun in 1980, have moved some way towards alleviating the dearth of information, but even these surveys fail to approach a comprehensive analysis. Yet the disagreement between the PEIs and the EC on where professional status begins further exacerbates attempts to analyse these factors. Certainly the PEIs in seeking an all-graduate occupation believe this will enhance their claims for professional status. Berthoud and Smith found that about six per cent were Oxbridge graduates and about forty per cent were from polytechnics, many of these obtaining their qualifications via sandwich courses (1981, pp.36-63). In 1985, however, the EITB claimed that 'almost three quarters of Britain's engineering graduates are produced by universities", the remaining possessing CNAA degrees (1986, p.19). Engineers have a greater tendency than any other group of professionals to have attended secondary modern schools, and have the lowest percentage of ex-public school pupils among their membership (Berthoud and Smith, op cit, pp.52-60; Gerstyle and Hutton, 1966, p.3).

Generally, in recent years about half the engineers who have responded to surveys were from middle-class backgrounds (using the Registrar General's classification) (Institute of Chemical Engineers, 1981, pp.11-51). The proportion for the country as a whole is about a third, whilst for the legal profession, by way of a contrast, it is nearly ninety per cent (ibid). It can be seen, therefore, that engineering tends to provide a ladder for the upwardly mobile, attracting people from a somewhat different social and educational background from the established professions and from politics and the civil service generally (10). Quite how important this is in terms of policy-making within the nuclear (or any other) sector it is almost impossible to measure, but it is possible to tentatively suggest that it does not facilitate communications between the groups involved in decision-making: the paramount importance of communication in the decision-making process being illustrated in the following five chapters.

It can be argued, therefore, that within the confines of the analysis of technical professionalism attempted in the previous chapter, engineering is not a fully professionalised occupation, it is a collectivity of occupations. But the sub-divisions into which it has fragmented have adopted the ideology of 'professionalism' and are attempting to pursue a policy of professionalism, with some such as the Civils, Mechanicals and Electricals, being more advanced than others. Although fragmented into its constituent disciplines, it must be stressed that a wider collective identity also exists, the fact of the pressure exerted by the PEIs and trade unions for higher status for engineers per se is a manifestation of this. One trade union official argued that:

> "(We) believe that engineering is a profession and that electrical engineers, whom we represent, are a part of that profession. Our members are very much aware of being professionals and are constantly pressing for higher status" .

But such a collective identity is only secondary: primary loyalty is to the sub-discipline and within that the factor of dual allegiance and the socialisation of engineers within the corporate hierarchies leads predominantly to individuals identifying with the interests of their employer.

This is reflected partly by the superfluity of the codes of ethics. Certainly the codes are in existence, but they are not enforced. Finniston wrote:

"We were not persuaded that the codes of conduct incorporated into the current registration system carry much weight or sanction in practice" (1980, para.5.35).

Not only could Finniston find no instances of an engineer being expelled from his professional institution for misconduct or incompetence (a situation repeated during the research for this study) but even if a member were expelled there would still be no legal sanction to prevent him from practising his expertise. The contrast here with the social professions is most stark, illustrating again the very different nature of the professionalism involved. A contrast further underlined by the belief of a member of the Finniston Inquiry that :

"...the function of an engineer is to make sure that the project with which he is associated works. If in achieving that he does his company proud, then he is serving his profession. The distinction between loyalty to an employer or to the profession is not a valid one" .

It is, as Prandy has argued (1966), noticeable that if an engineer is dismissed by his employer he is penalised far more than if he were expelled from his professional institution.

From the above discussion it can be argued that at its margins engineering merges both with the routine occupations and with the managerial strata, or techno-bureaucratic positions. It achieves its greatest unity and influence within its sub-disciplines (which in some instances become themselves quasi-professional) and within the structures of employing organisations, especially if those organisations perform the type of technical function found in the nuclear industry. In the following section of this chapter it is seen that although there are obvious contrasts many of these points are also applicable to science and tend to be confirmed by the research conducted for the substantive part of the thesis.

Scientists

The major distinction between science and engineering is the large research sector located within the former. Indeed, at its most basic there is an argument that engineering is parasitic upon the fruits of this function, applying them in the industrial sphere. Engineering is the routine application of science and it is propounded in the following chapters that with the dissemination of nuclear expertise from out of the AEA and into the routine world of electricity generation, it passed from the realm of science and into that of engineering. In the introduction to this chapter mention was made of the nature of science, that it is not a profession but a generic term describing a series of occupations, many of which are professionally organised. This suggests that at its boundary there is an inevitable blurring with the borders of engineering: it is these issues which are examined in more detail below.

There is no engineering equivalent to the prestigious Royal Society which heads the body of science in Britain. Engineers are, however, eligible to become Fellows of the Royal Society, an honour widely regarded with possessing greater status than the recently-instituted Fellow of Engineering (F.Eng.). From its foundation in 1660, the Society has enjoyed royal patronage and certainly in the first century of existence Science enjoyed greater prestige than Law or Medicine (Andrade, 1960). The society exhibits many of the hallmarks of a professional institution in that it is self-governing, being run by its Fellows who number about six hundred. Each year another twenty-five Fellows are elected from amongst the most eminent of current British scientists. The Society has a complex committee structure and this is used not only for internal organisation but also to establish and maintain close links with the wider national and international body of Science, with Government and also industry. It is one of the functions of the Society that it tenders expert advice to the Government when asked, and frequently when not if its members feel strongly about a subject affecting science. Its relationship with Government extends to its responsibility for dispensing some government-provided monies intended for the furtherance of research.

Despite its status, however, the Royal Society does not perform an array of supportive functions normally associated with a professional body. It is a unifying forum, representative not so much of scientists, but of science. It prefers to give its attention to questions of science per se, and not to become involved in issues of a professional nature such as terms of employment or levels of professional competence, leaving these to the professional institutions or learned societies. The professional representatives of the individual scientists are the professional scientific institutions (PSIs). In emulation of the engineers' attempts at overcoming occupational disunity they established the Council of Scientific and Technology Institutes (CSTI) in 1968. Its objectives were similar to those of the CEI, the pursuit of increased status through a continuing campaign of public awareness:

> "...to make known as widely as possible the part that science and technology play in the modern community (and to) enhance the contribution of science and technology to the well-being of every citizen" (CSTI and Mintech, 1968, p.1).

The CSTI and its members saw themselves as representing the professional scientists in a similar manner to that of the CEI and PEIs in regard to engineers.

Part of the reason for the establishment of the CSTI was to attempt to stem the flow of defections of registered scientists to what they saw as the more professionally successful PEIs. There is evidence to suggest that this move has not wholly prevented the desertions, at least one PSI secretary bemoaning that:

"The pressure is really on scientists...there is the danger that our members in the future will not consider it worth being a member of our institute, but will join engineering institutions, the IEE in particular".

It is the structure of industry which hastens the registration of scientists with the PEIs. In private industry, it is argued by the PSIs, 'the dividing line between scientists and engineers becomes non-existent' [11], and as a result a physicist is 'frequently indistinguishable from an engineer' [12]. Further-more, it is argued that

"If you ask senior personnel in large firms, such as ICI and GEC, how many physicists they employ, they could not tell you...this is the consequence of the fact that physicists and engineers work alongside each other and do the same job, becoming indistinguishable" [13]

a view reiterated by the officers of trade unions representing both scientists and engineers.

Even more than is the case with engineers (see above, this chapter), data regarding the distribution, work situation and sociology of scientists in Britain is poor. This paucity of information is in urgent need of remedial research, and for the present allows only the vaguest of inferences to be drawn regarding the occupation as a whole. As with engineers, it is not known how many scientists there are in Britain and a 1968 survey sent to its forty thousand members by the CSTI received only a forty per cent response rate. In 1971 the returns were even more meagre and the surveys were discontinued on a large scale. The EITB survey of 1985-86 counted over 18 thousand scientists and technologists, but as with engineers, those in managerial positions were not included.

From what evidence is available it appears that, as with engineering, the largest employment sector is private industry, where just over half the scientists are located [14]. Nationalised industries employ only about six per cent; universities about thirteen per cent; but central and local government (including the AEA and the armed forces) account for about a third. Less than one per cent of scientists appear to be self-employed. A further similarity with engineering can be seen in the method of career progression, with over half the survey respondents (the older members) claiming to be employed primarily as managers. The one major contrast with engineering unearthed by the surveys appears to be that the teaching and research sectors are much larger. Perhaps the most important point suggested by the sparse date is that in the private sector the overlap between scientists and engineers is such that when it is combined with the nature of dual allegiance, it makes more sense to view these people as technologists or technocrats, rather than the seemingly redundant labels of 'professional scientist' or 'engineer'. Their identity is defined more by their employer than by any vestigial allusion to the concept of professionalism per se, although as the analysis of Chapter Four seeks to illustrate, this concept retains some appeal. At the level of the individual, it can be argued, motivation becomes awash with both organisational and professional aims. This is particularly so in the public sector scientists.

It appears that in contrast to their private sector colleagues, public sector scientists enjoy status and privilege through emphasising the nature of their scientific calling. The analysis of the following three chapters tends to support the proposition that the more an organisation is dependent upon research the greater the prestige and influence of its scientists. Prandy argues:

"Scientists have always emphasised the importance of basic research, partly in order to maintain control over their own subjects, since clearly only they are competent to judge the value of such research" (1968, p.18).

There are obvious links in this view to professionalism's goals and it is apparent that employing organisations which support a work structure amenable to those goals will become the centres of professional excellence. In Science these are the universities and public research organisations. Dunleavy argues that since in Law and Medicine it is private practice which most epitomises the professional goals of self-regulation and work autonomy, these are the most prestigious professional sectors. In those terms, the universities can be viewed as a surrogate private practice for Science (1979, pp.19-35).

The repercussions in that for this study are important and are analysed in some detail in the next three chapters. As an organisation primarily devoted to research, the AEA is an example of an institution dominated by its scientists, their career structure as determined by the inherent sub-disciplines of the occupation, influence the goals and procedures of the Authority, within the broad parameters set by the original governmental aims. At least one former employee argued that this perception of their professional status decided the approach people took to their jobs:

"It was the career structure that influenced people's decisions...we had at Aldermaston science groups who prided themselves on being scientists and projected themselves, with a view to getting on in the world as first-class scientists".

It will be shown in the next chapter that this attitude was rampant in other parts of the AEA and was as a result of the influence of the status of the university sector, the division into sub-disciplines reflecting the method of organising Science in these institutions. The structure of Science outside of the private sector is therefore more apparent, particularly where novel or pure research is practised. Even in the parts of the AEA run along university lines (Gowing, 1974), however, science could not be construed as a united profession in the way such an interpretation could be applied to Law and Medicine (see the previous chapter and its discussion of this subject). Any sense of unity was limited to the employing organisations, or more usually, a sub-organisational team in which an individual represented his particular sub-discipline. As in engineering, professional organisation occurs at the sub-discipline level, that is where people describe themselves as 'chemists' or 'biologists' etcetera; a facet with obvious links to Kuhn's definitions (see Chapter 2 above). But, as it is attempted to show in future chapters, organisational loyalty intrudes upon this community spirit.

The problems of establishing social and educational back-grounds in science are a repeat of those encountered in engineering. Again sociological data is so restricted that the expression of any firm opinions would be foolish. Who's Who entries are really all one has to proceed with in regard to the professional elite and these show that forty-five per cent of the scientists on the AEA's Board between 1962 and 1982 were Oxbridge graduates. Yet since this is a sample of just twenty-two people it throws very little illumination upon the background of the scientific elite generally. Data on the vast majority of scientists is even less forthcoming than that available for engineering, although the CSTI surveys have tended to suggest a slightly greater emphasis upon a middle-class, grammar- or public-school background than that pertaining to engineering. Berthoud and Smith found that about fourteen per cent of scientists (six per cent of engineers) tended to be Oxbridge graduates (1981, pp.52-60) and that only about twenty per cent were polytechnic products. Generally, the need for more detailed research into this area must again be emphasised.

This section on science has suggested that, like engineering, science should be viewed as a collection of sub-disciplines, some of which are more professionalised than others according to the definitions established in Chapter 2. In contrast to engineers, scientists more nearly attain the goals of the professional ideology when they are situated within organisations devoted to research, thereby reflecting the most pure aspects of the scientific calling (this does not, of course refer to applied research). The collegiate nature of scientists is nurtured and then utilised in the centres of professional excellence found within the university and other liberal research establishments. Within private industry, however, the community is subordinated to the hierarchy and is subjected to a general lessening of professional identity, engineers and scientists merging into the technocracy, where the concept of professional ethics for example applies even less than in the public sector .

Conclusion

This chapter has attempted to analyse the occupations of engineering and science from the perspective of the definitions obtained in Chapter Two. It is fairly obvious that the techno-crats are of a different type of professionalism from that which pertains in Law and Medicine, their lack of a cognitive and structural unity and their position within the hierarchies of the public and private sectors ensuring that their aspirations for the type of professional autonomy achieved by the social professions would remain unfulfilled. Dunleavy's arguments that in a professionalised policy-system the stronger the unity of the community, the more likely it is that influence flows take place between the private and public sectors (1979, pp.12-15), would tend to suggest that in the case of science and engineering such flows would be less than in other occupations: the main influence of the ideology of professionalism being within the corporate hierarchies. If that is the case then it would seem that his further arguments regarding the nature of professionalism as ideological corporatism (1981, pp.3-15) would tend not to be fulfilled: namely he argues that under certain circumstances the interest corporatism model which typifies organisational bargaining recedes into the background. In its place ideological corporatism takes over to pursue the technical and professional objectives of unified professions, subverting the institutional barriers of the private and public sectors and intruding upon the political process.

Such a process, a typical product of the professionalisation of policy-making (Dunleavy, 1981) would not seem so feasible in the disjointed structures of the technical professions of the nuclear power policy-system. Just how extensive such professionalisation is in the nuclear industry is the subject of the following chapters, where it is argued that although the professional goals and structures of the engineers and scientists can subvert inter-organisational boundaries, the greatest influence of the technical professional is found within the hierarchies of their employing organisations. Professional structures may cause influence flows to take place across the public and private sectors, but as the discussion of the fragmented nature of engineering and science in this chapter has tended to suggest, the greatest impact of the professional ideology in the nuclear industry could possibly be when it combines with the organisational goals of the professional/managerial elite of the public sector organisations.

In order to examine this possibility and its ramifications the analysis turns, in the following chapters, to an exploration of the influence of the technocrats in shaping nuclear policy. The initial chapters have sought to explain the concepts of "professionalism" and "technical professionalism" and now these are to be used as the perspective from which to analyse the nuclear power decisions that have been taken by policy makers in Britain. In particular, it is attempted to show how the disjointed and fragmented structure of "science" and "engineering" outlined in this chapter is combined with the dual allegiance of individual technocrats to have a profound collective influence upon policy-making.

NOTES

1   See Chapter 1 for a critique of trait theory and Wilding, 1982.
2   Much of this was expressed in interviews.
3   Interview.
4   Ibid, and Recommendation of the Report, 1980
5   A point made forcefully by several interviewees, particularly those with a senior rank within the PEIs.
6   Various interviews. There are, however, no readily obtainable figures to fully substantiate this. It can be

argued that the membership and the unions themselves were playing the trade union card in an attempt to frighten the PEIs into action.

7. Sources: Prandy, 1965; Gerstyle and Hutton, 1966; Berthoud and Smith, 1981; CEI and Mintech surveys between 1965 and 1980.
8. Various interviews
9. Obtained by analysing Who's Who entries and comparing them with available survey data regarding the bulk of engineers.
10. For an assessment of the social background of civil servants see the Fulton Report (1968).
11. Ibid.
12. Ibid.
13. Ibid.
14. All these statistics are taken from the 'Survey of Professional Scientists' 1968 and 1971, Mintech (DTI), CSTI.

4 The professional structures I

Chapters 4 to 7 provide much of the body of the thesis devoted to exploring professional influence within the nuclear policy system. Chapters 4 and 5 should be read together as they are complementary accounts of the professional structures found in the nuclear industry. Beginning with a description of the Atomic Energy Authority (AEA), this chapter proceeds to map out the position and role of technocrats within that organisation, before continuing to analyse the Authority's relations with private industry and the political system. The initial conclusions lead into the content of Chapter 5 where a more detailed conclusion of the two chapters prepares the ground for Chapters 6 and 7 which concentrates upon technocratic influence in the reactor decisions.

The sequence of the chapter is conditioned by its position within the whole. The necessary theoretical discussions of the opening chapters paved the way for a practical discussion of "science" and "engineering" and it is now important to relate that to the nuclear industry per se in order to analyse technocratic influence. Because Chapters 2 and 3 argued that professional power was partly dependent upon occupational structure and position, a view reinforced by the theory of duel allegiance, it is logical to begin with a review of professional structures and a discussion of their importance; a sequence repeated in the next chapter. The points raised in that section are then related to the private industrial and the political spheres in order to ascertain the wider influence of technocrats. The latter concern providing a convenient lead into Chapter 5's account of the CEGB and its increase in political influence, an increase that has corresponded to a fall in the powers of the AEA.

What begins to clearly emerge is the importance and strength of "politics" within the British system, unlike the single-minded and pragmatic approach of (say) the French towards nuclear power. Political "reality" as it intrudes upon technical "rationality" is seen to be the deciding factor at all levels of decision-making. It is this powerful policy determinant that imbues every major organisation and the decisions of that organisation, those able to manipulate it "win" their decisions. It is the realisation of this reality and the skill with which they have performed that have led the technocrats of the CEGB to supplant those of the AEA as the senior members in the coalition of organisations comprising the nuclear industry.

The Atomic Energy Authority

The history of the AEA is a catalogue of almost continual change in both its structure and influence. The change in the influence of the Authority can be traced through its effect upon the political policy-makers of Whitehall and Westminster, and the technical decision-makers of the electricity supply industry and their component manufacturers in private industry. Essentially, the entropy of the Authority reflects the declining importance of pure research and applied science to the success of the nuclear power programmes as the more mundane problems of established engineering practice, management and workforce relations and manufacturing difficulties, became the limiting factors in the policy process. It is these which decided whether the policies eventually implemented closely reflect the intentions of those who initiated them.

There have been at least four periods of major reform, the most radical being in 1971 when the AEA lost the Atomic Weapons Research Establishment at Aldermaston to the Ministry of Defence; its fuel fabrication and reprocessing facilities (the old Production Group) to the newly-formed British Nuclear Fuels Limited, which henceforth was to be a nationalised industry with all the shares held by the Authority on behalf of the Secretary of State for Energy; and also its radioisotope centre at Amersham. The latter was to be one of the first public industries propelled into the private sector by Mrs. Thatcher's Government in 1980, having spent the period from 1971 as a highly profitable public corporation with its shares also held by the AEA on behalf of the Secretary of State. Chapter 8 details the evolution of BNFL from its origin in the Production Group to a situation of commercial success as a publicly-owned company. The other major reforms were the establishment of the Development Group (1959); the birth of the Reactor Group (1961); and the reorganisations of 1965.

Gowing (1974), Williams (1980) and Ince (1982) all provide explanations that illustrate the considerable extent to which the prevailing circumstances of post-war Britain dictated the structure of the Authority. The overriding atomic priority of the Attlee and

Churchill Governments was the design, testing and production of British nuclear weapons, and the structures that originated from within the old Ministry of Supply reflected this aim and reflected also the constraints of a country which found itself in straitened circumstances. A further legacy of those years was to be obsessive secrecy which necessarily had to surround atomic weapons projects but was also carried over de facto into the civil side (Gowing 1974, Ch.16), a factor which at least one author charges led to a lack of public accountability and discussion that could have alerted decision makers to the problems encountered by the ESI in the nuclear power programmes (Williams 1980, Ch.12). The basic research necessary for the design and production of nuclear weapons and civil power was carried out at Harwell in Oxfordshire.

Built on the site of a wartime airfield, Harwell was entrusted to Sir John Cockcroft. Gowing details the activities of the early years, emphasising the liberal regime of Cockcroft who was determined to pursue a University style of procedure. Individual scientists were encouraged to research a plethora of interesting, even idiosyncratic areas; indeed such a regime was felt by Cockcroft to be a necessary prerequisite to attracting the best talent out of the Universities, where they enjoyed considerable academic freedom, and into the government sector (Gowing, 1974, Ch.18). The production side of the weapons programme, based in the northern factories at Risley, Windscale and Capenhurst, followed an entirely different regime. Under the leadership of Sir Christopher (later Lord) Hinton, this sector was organised along the lines of a commercial enterprise or at least as close as it was possible to achieve without the incentive of the need to make a profit. The groups controlled by Hinton were mainly staffed by practical engineers and they were subject to an industrial and commercial management style, developed during his pre-war stint as a senior manager at ICI. Hinton pioneered cost centres and accountable management on a large scale in the UK and he strove single-mindedly to achieve his government-determined goals on time and to budget. Activities not specifically designed to facilitate these targets were rigorously curtailed (Gowing, 1974, Ch.21,22,17). It was in these years that ties between the Production Group's predecessors and certain sectors of private industry grew close; however, attempts to emulate the American experience and pass the larger share of the design and construction work onto the private sector, as Burn argues should have been done (1978,Part One), were simply not feasible in Britain. The sheer scale of the programme meant that British capitalism was incapable of approaching it: indeed it actively sought to avoid performing in anything other than a sub-contracting capacity for individual components, preferring to follow the specific instructions of Hinton, his managers and draughtsmen (although industry was not above 'poaching' those same draughtsmen after Hinton had trained them) (Gowing, 1974, Ch.17).

Finally there was the Weapons Research Establishment at Aldermaston and its associate outposts in the old wartime ordnance factories. In certain places, such as Windscale, there were (and still are) close links with the research and production sectors: the predominant mood, however, was that which was closely modelled on the ordnance and military background of the staff, with the exception of the head of the group, Sir William (later Lord) Penney. That is, research was of the purely instrumental kind, ruthlessly aimed at the attainment of specific goals; costs were of a secondary concern and management was structured and styled to achieve the one aim of building nuclear weapons as quickly as possible (Gowing, 1974, Ch.14,16,23,24).

This tripartite division was repeated in the Authority when it was established under the 1954 Act. The production section became known as the Industrial Group, before becoming the Production Group less than a decade later and then a decade after that being siphoned off again to form the basis of BNFL. The first two chairmen were powerful Whitehall figures, Lord Plowden and Sir Roger Makins (later Lord Sherfield) a Treasury mandarin. This reflected both the civil service background of the AEA and also its tremendous prestige. Lord Plowden in particular was to be able to take a cavalier attitude towards costs, something that was not easily forgiven by the Treasury (see Chapter 8 below). From the outset the Authority had a high level of status and its constitutional position, as set out in the various sections of the Act, was as the Government's adviser on nuclear energy. Thus, not only was the AEA to conduct nuclear research and production; it was to be the Government's expert on the subject, a potential conflict of roles that is discussed below. From 1964 the Chairmen of the AEA were to be scientists, first Penney, then Sir John Hill and W. Marshall and Sir Peter Hirsch; only in 1984 was there a return to a Chairman with a civil service background in the appointment of Arnold Allen, a long-standing member of the Authority and a former Treasury high-flyer. There was in the early years a federal structure, each Group enjoyed a considerable degree of autonomy and functioned independently of the other Groups in several areas, although of course in other areas there had to be considerable integration, examples of the latter being research into fuel fabrication and the detailed metallurgical research that was involved (Williams, 1980, pp.50-53). The weapons research and development was the agent that shaped the structure of the AEA and the independent groups that emerged were as a result of the early policy decisions.

Other factors that had a profound contribution upon the structure of the Authority resulted from Britain's changed position in the world, particularly in relation to the United States. From its position of financial and military dependence the United Kingdom was in no position to demand the abrogation of the McMahon Act that suddenly and absolutely ended the wartime military collaboration of Britain and the United States of America in nuclear matters. British illusions about being a victorious world power demanded, however, that the UK should seek to retain a seat at the top table of nations and after Hiroshima that meant acquiring atomic weapons. The problems of achieving any kind of nuclear potential, civil or military, in the face of American non-cooperation were formidable (Gowing, 1974; Williams, 1980, Ch.1). A senior metallurgist with the Authority during its early years, who worked at Harwell and in the North for Hinton, recalls that the decision to pursue the gas-cooled option instead of the water-cooled method preferred by the Americans in their military and civil reactors was as a result of the American intransigence over collaboration. Repeated British pleas for joint ventures following the successful wartime cooperation were either ignored or bluntly refused in the early years. A member of Hinton's team recalled:

> "People speak as though we have made great choices; we haven't. To start with we had no options actually, none at all."

The main problem the American isolation caused was in the lack of raw materials, as much of the basic research was contained in the experiences and notes of the British members of the wartime teams. One of them argued:

> "When I was in atomic energy in 1945 we were considering PWRs; we went the Magnox way because we could not get enough enriched uranium 235. So we had to build something that could burn ordinary uranium."

The decisions of the 1940s and 1950s were dictated by a mixture of national hubris and circumstances: politicians decided that Britain should possess nuclear weapons, and organisations established to obtain them threw together thousands of people who were in possession of considerable talent which they used to urge the realisation of the civil as well as the military potential of nuclear power (Gowing, 1974). Powerful figures close to Churchill followed Lord Cherwell in pressing for the organisation to be hived out of the civil service in order to facilitate flexibility of action unencumbered by Whitehall bureaucracy (AEA 1979 AHO/4). These basic decisions were then implemented, but the manner of their implementation was decided by the factors outlined above. It was the technical experts interpreting and implementing their brief within the climate of secrecy and national isolation who made the crucial decisions regarding the types of programmes followed and the method of operation in those early years. The result has been a history characterised by what Brian Wynne has described as a policy-making process of incremental and ad hoc decisions (1982, 1978). Thus, decisions taken in the early 1950s led inexorably towards certain conclusions being formed in the 1960s and 1970s, a member of Hinton's team argued that:

> "once we had (Magnox) it was natural to go into AGRs, even when we had enriched uranium".

That is, the investment in resources made in the gas-cooled technology dictated that to repudiate that technology in favour of the American system was increasingly difficult. Yet there were decisions taken to pursue different systems (see Chapter Five for an analysis of the SGHWR and the decision by David Howell to adopt the PWR); this is the ad hoc part of the policy-making process, decisions made by politicians that do not, at first glance, follow an incremental pattern.

It can be argued that the divisions between the different Groups in the Authority represented a division between scientists and engineers. Harwell under Cockcroft was (and still is) a prestigious scientific establishment with the different scientific disciplines organised and operating along similar lines to departments in a university faculty. It corresponded to the situation of professional autonomy desired by professional and professionalising occupations as outlined in the previous chapters. In particular it functioned almost as the surrogate private sector which Dunleavy argues is the role performed by the university sector in those occupations wholly or predominantly subsumed within the public sector and therefore lacking a prestigious private practice group (Dunleavy, 1979,pp.19-35). But the degree of occupational freedom accorded to the scientists should not be over-emphasised. However much this may have differed from the production parts of the Authority, Harwell still had targets to achieve and still had to perform basic and applied research as and when it was required for the weapons programme and then increasingly (exclusively after 1971) for the civil programmes. Another area where Harwell performed a university-like function was and is in the publication of scientific papers, the holding of seminars and the general diffusion of knowledge. This is dealt with in more detail below, but essentially Harwell was an important school where many of those in the ESI and manufacturing industry were trained. The Industrial Group led by Hinton was the preserve of the engineers. There were several types of engineering occupations welded together into integrated teams, the work contributing to the development of the chemical or production engineering and metallurgists in the UK. Although the work was concerned with the implementation of the fruits of the scientists' endeavours in the early years and in the conversion of theory into engineering practice, Hinton's teams also carried out a large amount of research (or practical science) themselves, indeed Hinton was a skilled researcher as well as a consummate engineer and gifted manager (Gowing, 1974, Ch. 14, 21, 22; Williams, 1980, Ch. 1, 2).

This division between the North and South, between scientists (Cockcroft) and engineers (Hinton) was not a conflict, which implies a positive state of opposition, but more the practical result of different sets of world views. As Chapters Two and Three show, there is a variance between the occupational ethos of scientists and engineers, the generic term science covering a diaspora of occupations, some of which show a markedly higher rate of professional autonomy than others. Engineers, although they also belong to a plethora of different occupations, possess a contrasting ethos in the sense of a somewhat lower professional attachment in most branches of the "profession", indeed as is shown above, the phenomenon of dual loyalty and bureaucratic patronage means that most of the engineers are "professionalising" rather than "professional", and that many are only quasi-professionals, if that. These cognitive differences are then reflected within the policy-system of the AEA, with the research sections being science-dominated and therefore naturally reflecting the prevailing norms of the scientists, whilst the production groups were engineering-dominated, the past tense reflecting the drastic changes that have led to he virtual removal of engineering from the ambit of the Authority in all but an ancillary capacity. The evidence to support such a view of the divisions within the authority is well documented by Gowing, but there is further confirmation in the evidence presented by Penney to the House of

Commons Select Committee on Science and Technology in 1967. In this, Penney, a scientist, moved away from the integrative approach of Hinton and instead reflected the Harwell/scientist view of the problems associated with the research into civil nuclear energy. He argued that problems are:

> "...much more easily and sensibly grouped as disciplines than they are as projects." (HoC 381-1, evidence 2/3/1967)

He proceeded to illustrate how he preferred to group the problem of corrosion in reactors, all reactors, as a problem for the chemists than as one for the teams engaged specifically upon the Magnox, AGR and HTR etc. Although some areas, such as the field or radiation damage, were carefully integrated with reactor development, through co-ordination committees:

> "...the research people (have) the opportunity to look well ahead and therefore in a sense set themselves some of the problems." (Ibid)

The integration and centralisation of the work of the Authority became easier after the reorganisations following the establishment of BNFL, Amersham International, and the relocation of Aldermaston to the Ministry of Defence. This reduction in the size and scope of the Authority, by removing the majority of engineers, left mainly a science-based research organisation. Development on all but the fast-breeder and fusion reactors is something that lies in the past or with other organisations (Serby, 1984, p.5); the Authority has now an almost purely research role.

Generally, Penney's preference for a discipline-by-discipline type of approach is not one that has prevailed throughout all the research parts of the Authority. As is shown in the previous chapter and in Chapter 8, the need to approach both research and production via a team effort has served (and does serve) to reduce the parochialism of the technocrats. Allen, when Chairman, sought to promote an integrative effort within the Authority as a whole, but Hill also brought his experiences, as head of Production, to bear in attempting to instil a more corporate approach within the AEA, something he facilitated with the removal of the Production Group altogether. The Authority does not promote on grounds of discipline, but all scientists are lumped together as group A and the rest are Group B. Because it is a technical body and because many of the policy decisions require a technical knowledge and background, the top jobs in the AEA have traditionally been held by the technically qualified, and the Authority has nurtured the belief in these people that they are "Authority people", not "professional people" first. There is no sudden 'V' in the road where they go into management and cease to be scientists or where they continue to do research; the gradual movement from being a technical operative to being a technocrat is a complex and lengthy process of training and socialisation or goal assimilation.

Initially it can be argued that the scientists and engineers had been influential in shaping policy within the AEA, within the parameters set by the politicians. That is, the politicians originally outlined the requirement of atomic weapons, but it was obviously up to the technocrats to apply their technical expertise in interpreting the best way of obtaining this goal. The divisions within the Authority between production and research, engineering and science, were natural concomitants of the contrasting occupational ethos. Although it led Hinton to complain bitterly at times that the scientists at Harwell were too busy pursuing their personal research goals instead of finding solutions to his problems on the production side (Gowing, 1974, chs.18, 21, 22), provoking him to establish his own research facilities at Risley (Gowing op.cit. and various interviews) the relations were generally complementary. A former AEA executive argued this view, noting that he did not think were was friction between scientists and engineers: "They have different viewpoints and therefore necessarily approach things from a different perspective". Certainly as the authority gradually achieved its goals in first the military and then the civil side of its operations, the evolution of the organisation led to sections evolving out altogether from the AEA. In many ways a comparison can be made with the Home Office, in that a body was established to oversee a series of functions, which as they develop and acquire a separate identity sought a separate existence, reasons of flexibility, management control and efficiency made it desirable and possible to hive them off.

As to the actual methods of attaining the goals of a British nuclear deterrent and then a civil nuclear capacity, the practical restraints of American non-cooperation, lack of raw materials and enrichment facilities, combined to unite with the stringent safety criteria necessary in such a small and densely populated country, to select the gas-cooled, graphite-moderated reactors that Britain has developed a course of action for which the technocrats were responsible certainly, but which in the circumstances, if not entirely unavoidable, was at least fairly predictable or self-selecting. The role of the AEA's experts in influencing policy is further illustrated when the relations between the Authority and the manufacturing sector and between the Authority and the political policy-makers is analysed. It is the former that is first explored, followed by an analysis of the latter, a sequence that leads into the discussions of the next chapter where the very different technocrat/politician relationship of the CEGB to Whitehall is analysed.

Relations with Private Industry

Duncan Burn published a scathing attack upon the AEA and the way in which it has dominated the nuclear policy-making of the civil programmes to the extent of designing the reactors chosen by Britain (1978). His critique is that the Authority should have limited itself to a narrow brief of being the Government's chief adviser without indulging in its own detailed industrial-scale research and development this, he argued, should have been passed over to the private sector in consultation with the electricity supply industry as soon as the basic research work was completed. The scale and type of operation carried out by the AEA was, he argued, a factor in the Authority's advice to the Government which must obviously have been biased given the AEA's own deep and continuing involvement in the nuclear programme. There was, he said, "a complete lack of a competitive background" and this was caused by the lack of private industry's involvement. Furthermore, this lack:

> "...compounded the threefold deficiency in development, skill and experience. The AEA developers had nothing with which to compare their work: no standards save their own image. As a result of the narrow front, there was no possibility of internal comparisons within the Authority. As a result of the AEA monopoly there was no domestic comparison from outside of the Authority. There was no competition from overseas (except in export markets where Britain invariably lost). There was no competition at all so long as the United States was not offering a competitive reactor...and...competition could even then be heavily weighted. (1978, pp.159)

This charge is refuted by Gowing and by the Authority who argue that the Government "became sole entrepreneur not by expressed intention but because private industry refused the role" (Gowing, 1978, p.155). From 1945, the initial desire within the Ministry of Supply had been to use private industry to produce the requisite nuclear products and provide all the backup services, "just as the United States project had done" (ibid.). Many of the smaller firms were not capable of undertaking the work required of them whilst the chemicals giant ICI, which had been closely linked to nuclear energy R. and D. from the early years of the war was loath to become involved even after an initial agreement in principle. There were two reasons for this. The practical one was that the strict secrecy which would be necessary would require a restructuring of existing ICI methods of work and it would prove difficult to recruit a large enough supply of manpower. The second and political reason for ICI's reluctance to become involved was that it feared nationalisation by the Attlee Government and it felt that a high profile as a "commanding height" in the economy, especially in such a novel and important field, would encourage calls for its nationalisation (ibid, p.157). Hinton was keen to recruit leading industrialists like Kearton to work with him at Risley, but even this was to prove difficult. Kearton, for example, was offered a lucrative contract with Courtaulds, which he accepted, and even lower down the scale the remuneration the Ministry allowed Hinton to offer to middle and senior managers was often considerably less than that offered by private enterprise.

It became clear that:

> "...Hinton and his organisation became the entrepreneurs for all the plants in the first stage of the atomic energy project, not for reasons of political philosophy but because the private firms approached would not touch the job at any price." (ibid. p.160).

To point out as Burn has done that the United States was successful due to the inherent logic of using profit-orientated, market-disciplined private firms, misses the very real contrast between the strength of American industry and, as Gowing points out, the weakness of Britain in this field, having only the one firm, ICI, which could compare to the half dozen or so American companies. Although the three groups involved in the early years wished to "make the greatest possible use of industry" all of them "were disappointed that industry did not give more help" (ibid, p.193). Instead, Hinton and Cockcroft were forced to use their own scientists and engineers to design and manage the building of the early factories and research laboratories. Hinton in particular was anxious to involve industry but had to resort to his own devices in building up a complex industrial network in the North that corresponded to the creation of a new industry which (with Harwell and Aldermaston) by 1962 was employing over 2,600 professional scientists and engineers and over 20,000 industrial-grade staff (AEA Annual Report 1962/3). This was a large slice of the country's skilled manpower and Williams has argued that when total figures are taken into account, the AEA in fact grew to a peak of 41,000 in 1961 and that this threw British energy R. and D. strategy "out of balance" and had adverse effects upon "British technology as well" (1980, p.328); a point returned to in latter chapters but related to Burn's critique.

From the outset, however, the Authority and particularly Hinton, with his bias towards the private sector, sought to involve the private firms as closely as they would allow themselves to get. This meant that nearly all of the construction work (and much else) was contracted out, the contractors and sub-contractors working under the direction of the AEA managers. By 1964/5 over 17,800 contracts were placed annually with private firms, worth a total of over £43,100,000 (AEA Annual Report 1964/5, pp.7-9). This was a considerable slice of the public sector funding and was therefore an important source of private sector support and profit, a factor noted by Benn in his evidence to the House of Commons Select Committee on Energy in 1980, although he referred to the whole of the policy system including the power station construction programme (Vol.11, HC 11 4-11, pp.394-434; see also Benn, 1984). It can therefore be argued that in the initial years and until comparatively recently (see below the discussion on the CEGB) the scientists and engineers of the Authority not only decided the policy of the Authority, but also structured the parameters within which private industry engaged in the nuclear programmes. This is an argument that is shown to be even more tenable below, in the discussions on the role of politicians and civil servants and in the next chapter where the decisions regarding the reactor choices are analysed.

One of the most important connections between the AEA and private industry and one that was to have significant repercussions throughout the period covered by this study, was the decision to establish the consortia to build the power stations. This aspect is analysed in more detail in later chapters where the various reactor decisions are dealt with: however, the relationship of the AEA's scientists and engineers to those of the private firms in the consortia was initially one of tutor and pupil, with the Authority's experts being the sole repository of the knowledge necessary for the effective and safe construction of nuclear machinery. From the early 1950s, when the Government first decided to proceed with a civil nuclear programme, the Authority had provided advanced facilities for the education and training of engineers in the CEGB and the firms that would manufacture the stations. By 1964 the Authority was able to claim that its relationship with industry was one of "close partnership" (Annual Report, 1964/5, p.59), although just three years later one of the consortia, the Nuclear Power Group, were to claim that far from being a partner they were:

> "...dominated on the one hand by the AEA who have a monopoly in reactor development and fuel manufacture, and on the other by the CEGB who are virtually a monopoly customer." (SCST, op.cit. Memorandum submitted by TNPG)

To a certain extent the criticism by TNPG was unfair to the Authority for by 1967 the AEA was already hoping to reduce its design functions by hiving off those aspects to private industry. Indeed, Penney told the Commons SCST that he could see no sense in having a strong design and construction organisation within the AEA which then passes its knowledge onto the consortia who then use that knowledge to do their own designs; he hoped instead to have just one D and C organisation and for that to be substantially run by the private firms, although with an Authority input (evidence, HoC. SCST, 9/3/67). Penney's successor as Chairman of the Authority, Sir John Hill, continued this policy, pushing hard for the requisite mergers, the consortia themselves were in favour but the CEGB under Stanley Brown were opposed and thwarted the scheme. Such an organisation was eventually to emerge in 1973 in the shape of the NNC.

The authority followed is premise of close links to industry even before the 1955 White Paper (Cmnd.9389, Feb. 1955), establishing a reactor school at Harwell in 1954 for the training of design teams from industry (Williams, 1980, p.27 and various AEA reports). Although Hill was later to advocate strongly a rationalisation of the industry, in the early 1950s it was the AEA itself which, adhering as ever to the idea of commercial discipline, advocated that "the most effective industrial arrangement for developing nuclear power would be to have several groups" and that each of these would be "wholly capable of executing the complete or turnkey contract for a nuclear power station" (Williams, 1980, p.28). By 1956 there were five consortia formed, each seeking to tender for what promised to be a lucrative nuclear future. Each consortium was to be built around a boilermaker and a manufacturer of turbine generators, plus (usually) a civil engineering firm. It became clear that there was never going to be enough work in the domestic market for all these firms and that unless they could export there would have to be a severe rationalisation (for how this came about see Williams, 1980, ch.8; Ince, 1982, ch.4). Apart from just two of the early reactors there were no British exports and the cut-back in the nuclear programme in 1960 (Cmnd. 1083, June 1960) underlined the need for a restructuring. Such a restructuring led in 1973 to the emergence of just one consortium, NNC, and that had a large Government stake in it, held on behalf of the Secretary of State for Energy by the AEA. The decision by the Conservative Government in 1979 to build a programme of PWRs led in 1982 to even further changes to the structure of NNC (Ince 1982, ch.2; AEA Annual Report, 1982/3). A part of the Authority's D and C function did eventually pass to the private sector, but the type of root and branch reorganisation advocated by Hill from the late 1960s came about only gradually, so that it took until the early 1980s for the gradual entropy of the AEA to reach a stage where all of its commercial teams had left, with the exception of work on the Fast Breeder at Dounreay in Scotland.

The strange ambivalence of the AEA towards the commercial sector has existed throughout its lifetime. As was shown above, there was the division in the early years between the commercially trained and biased Hinton with his Northern factories and the scientists in the South with their traditions of Government service and/or university-based pure research. This contradiction manifested itself on several occasions, with Penney even telling the Commons Committee in 1967:

> "We do have commercial interests. We do manufacture. We do have a very complete cost consciousness. We do integrate all these ideas right the way down to our research. I think if you divorce research and so-called development from industrial application and sales, a terrible gap can open. But I would defend the Authority as not being that kind of animal."
> (Evidence 2/3/67, HoC. 381-1)

This was the sort of sentiment that the MPs wished to hear and it was well received. That later events were to suggest Penney's interpretation was a little off the mark was not something to intrude upon the Committee's complacency. The subsequent remarks were not even challenged when, taking him at his word about commercial interest, the committee asked about return on capital and Penney replied: "We are not wholly a commercial industrial business, not at all". (ibid).

Thus when it suited them, the AEA were able to claim either scientific/research needs, especially when seeking additional funding or, when seeking greater autonomy, they would claim "commercial freedom" to act in an independent manner. The latter is something that is analysed in detail in Chapter 8 below with regard to the evolution of BNFL from out of the Production Group of the AEA. Certainly those within the Authority who began their careers with it in the North either under Hinton or his successor, Sir Leonard Owen, were imbued with a desire for commercial behaviour. Hill in particular wanted the widest possible dissemination of atomic knowledge:

> "...it is a right and proper way for experience to be transformed from a pioneering organisation to an organisation picking this up in the ordinary commercial, industrial sense".
> (ibid, 8/6/67)

He was understandably aggrieved when after he had retired he felt that he had been unjustly accused of being against the private sector; this is in fact contrary to the evidence of a career that saw the commercial transformations of large parts of the quasi-governmental sector and a movement that can be described as being "statism" to "privatisation" (see Chapter 8). Hill wished to transfer that which was commercially viable to the private sector if possible. He felt that the Amersham procedure, whereby the radioisotope part of the Authority was separated off, became Amersham International with the shares held by the Authority, and ultimately privatised, was the correct procedure.

This theme of commercialisation and its relationship to the role of the Authority as a research organisation is one that is returned to in more detail below as it is an important aspect for the understanding of the role of scientists and engineers. The interpretation placed upon it in this book is that when viewed within the context of the professional ideology it can be argued that the quest for commercial viability is another factor in the claims for occupational autonomy expounded by the technocrats. That is, pure science is closely akin to pure professionalism, yet the production groups were no longer engaged upon pure science: on the contrary, they were moving into the arena of mundane occupations, with the result that they became vulnerable to the same commercial and political considerations of other groups of workers. The pursuit of commercial status within the framework of the liberal democratic state was one way the technocrats sought to insulate themselves. They did this by allying with the goals of the decision-makers and thereby seeking to use their technical expertise in an influential manner by both putting it at the disposal of the decision-makers in order to help them make more "rational" decisions, and thereby at the same time actively aiding the construction of the outcome. The gains hoped for were the professional ones of autonomy and influence.

In educating private industry, the experts within the AEA initially ensured that they maintained control of their product, a vital factor in the process of professionalisation, as the work of Larson (1977), Wilding (1982) and Johnson (1977), reviewed in the previous chapters, showed. This led to the nuclear scientists and engineers enjoying tremendous prestige and being considered the professional leaders of their respective disciplines, an aspect duly noted by authors as disparate as Sweet (1982) and Williams (1980). The links with industry remain, but the prestige is much diminished and therefore so are the prestige and influence of the Authority itself, a factor marked by the appointment of non-scientist (Arnold Allen) as the Chairman, and one moreover who obviously lacked the political clout of Plowden or Sherfield. The downgrading of the AEA was further marked at the end of 1986 when Allen was replaced as Chairman by a CEGB engineer. In the very act of disseminating the knowledge it had acquired, the AEA sowed the seeds of the eclipse of its influence, in that although such an educative role initially added to its power and professional status, other organisations being dependent upon it and its experts, the very process of dissemination meant that eventually the knowledge ceased to be the preserve of AEA experts and became mundane, common property.

This would not have ended the AEA's dominance had it continued to lead in the field of commercial applications, but the extent to which it could open new areas of research and exploit them diminished as commercial exploitation made the product an engineering problem. The scientists regrouped around the more esoteric pursuits of the Fast Reactor and Fusion. The only way in which the teams responsible for the increasingly mundane applications of professional knowledge could retain their status and control was to follow the work into the commercial sector and convert their goal of autonomy. That is, professional expertise and goals led to the transformation of the goals of the organisation to which the technocrats belonged, this then of-course acted in turn upon the experts themselves in a dialectical process. In a sense this maintenance of control over the routine application of knowledge is the sort of activity pursued by medical doctors and lawyers (cf. Wilding 1982), but the method and type of control exercised by technical experts, although perhaps modelled upon the social professions must, for the reasons outlined in Chapters 2 and 3, be rather different. The most outstanding example of these points is the evolution of BNFL from out of the Production Group of the AEA. Both BNFL and the Authority still seek to lead the evolution of private industry in nuclear matters; evidence to the House of Commons Select Committee on Energy showed the depth of the AEA's concern for manufacturing industry, the Authority interpreting their role as "being to ensure" that there is:

"...a sound technical and engineering base to support the national nuclear power programme and to manage their research and development such that this objective is achieved". (Vol.11, HC 114-11 p.90)

The Authority also feels, in common with the ESI, that a continuing programme of power station construction is a necessary prerequisite to a healthy manufacturing base and that this is a strategic necessity if Britain is not to become dependent upon imports.

Finally, part of the concern for private firms and the communities in which they are situated is part of the Authority's continuing public relations exercise; it is pointless to retain a viable industry if the political opposition to nuclear power becomes irresistible. One manifestation of this concern is the continuing practice of both the AEA and BNFL to recruit and train more apprentices each year than they need; the excess are then free to return to their communities and contribute their skills. This is an important consideration in isolated and rural communities such as the Caithness peninsula where Dounreay and the Fast Reactor experiments are situated (Atom, 324, April 1985). But relations with private industry are but one facet of the policy process. Relations with the political actors is of vital import and it is a consideration of the AEA's role here that is now turned to.

THE A.E.A.

Whitehall and Westminster

Formally and constitutionally, the AEA is the Government's adviser on nuclear matters. This section is concerned with the formal and actual relationships between the Authority's technical experts and the political process. The formal role of the Authority as the Government's chief nuclear adviser is enshrined within the 1954 Act that established the Authority in its legal form. The organisation's members are appointed by the responsible minister and are accountable to and through him, in the same way as any other nationalised industry, although the AEA's history within the Ministry of Supply makes comparisons with nationalised industries difficult, the Authority in essence being a hybrid; not quite part of the Government's Whitehall machinery, but not quite separate either. Initially under the aegis of the Prime Minister, then the Minister of Science, the Authority moved to the Ministry of Technology in 1964/5 (MinTech) under Frank Cousins. In 1971 it moved to Trade and Industry before ending up in its present location at the Department of Energy in 1973/4.

For much of the period of its existence the AEA has been under the supervision of a different sponsoring department from that of the CEGB, one of the factors hat have exacerbated coordination and communication between the two organisations. From 1969 both the AEA and CEGB were the responsibility of first the MinTech, then the DTI, then the Department of Energy. In 1955 the Atomic Energy Office was set up in Whitehall; its purpose was to serve the minister responsible for nuclear energy. It was always a small office, rarely numbering more than half a dozen civil servants, being absorbed into Mintech when it was formed. The successor to the office is the Branch 3 of the Atomic Energy Division, which is in the Coal, Electricity, Atomic Energy, nationalised Industries-Policy section. A section is headed by a Deputy secretary, a division by an under-secretary and the branch by an assistant secretary.

The Chairman of the AEA is the Minister's adviser and as such he has direct access to him; the AEA as a whole is the adviser of the sponsoring department and this means that the advice is channelled through the structure outlined above. In essence, the Authority likes to view itself, in the words of a senior manager, as providing the Government of the day with options; "its fundamental role is that of the nuclear servant". That this viewpoint is open to considerable reinterpretation is made clear below. Certainly, that may be the post-1979 situation, but from 1954 until comparatively recent times, the Authority not only presented options to the Government of the day, it also conducted some quite considerable campaigns, lobbying in order to secure the option it preferred.

The structure outlined above is a fairly rigid system and successive Governments have preferred to limit themselves to the style and presentation of the policy advice that resulted, advice obviously dominated by the Authority line; indeed, as is shown below and in the remaining chapters, a recurring criticism of the policies pursued by governments emphasised the lack of independent government assessment of AEA advice. Successive Governments preferred to accept what has been channelled to them directly from the AEA or indirectly via the sponsoring department, arguing that an independent departmental expertise to review the advice of the AEA was needlessly lavish since it was the constitutional position of the AEA to advise on nuclear matters and not that of another body. An example of this is the

Government's reply to the criticisms of the Select Committee on Energy in its Green Paper "Nuclear Power" (Cmnd.8317) where it rejects the committee's comments attacking the lack of independent expert review in the DEn, referring to this as a "needless duplication" (pp.19-20).

Yet studies (cf. Gray and Jenkins, 1985) concerned with the policy-making processes of Whitehall suggest that considerations of formal structures can be placed on one side; it is the way in which those presenting the policy options are perceived by the policy-makers that really decides which option is pursued and the manner of its implementation. This is due to the informal structures and personal contacts that contribute to a special Whitehall ethos in which personal appraisals of people, rather than of policies or recommendations, are the deciding factor. The importance of the Whitehall ethos has been widely accepted ever since the pioneering work of Heclo and Wildavsky was published (1974) and such a view has been refined and contributed to by more recent studies (Ashford, 1981; Gray and Jenkins, 1983, 1985; Zifcack, 1984). It can be strongly argued from the evidence presented throughout this thesis and elsewhere that the politicians and civil servants who are responsible for deciding upon the options followed are not policy-makers in the sense of initiating and formulating policies and as such the references to them in standard texts are misleading. As will become clear in the following sections of this chapter and in the following chapters, politicians and civil servants are mostly reactive and not initiative; initiation and formulation in the policy process are functions performed by the technically expert within the policy community, and therefore politicians and civil servants are not 'policy makers', but are de facto policy arbiters, choosing between the various options put before them.

The degree to which the technical dependence of the politicians and civil servants was expressed as deference to the policy recommendations of the technocrats is glimpsed at in a 1965 Commons reply of the then Minister responsible, Maudling, to some minor criticism of the AEA by Arthur Palmer. Maudling told Palmer that the Government felt it enjoyed "the advantage of very comprehensive advice from a remarkably efficient range of people" and did not wish to interfere with the accepted programmes than under way (quoted in Williams, 1980, p.83). Palmer himself in 1965 was, like most politicians, fulsome in his praise of the nuclear knights and "yielded to none" in his admiration for them (ibid). The Authority's scientists enjoyed enormous prestige and access to governments throughout the 1960s: they had something of a free run, which in later years was to be the focus of criticisism.

A growing band of politicians (mainly in the Commons SCST) began to share unease at the influence of the AEA, an influence enshrined by its formal role and reinforced by the technical inexpertise of Whitehall and the clear uncritical deference to the Authority shown by ministers and civil servants alike. Palmer (as a member of the SCST) had consistently argued that the civil nuclear energy functions of the AEA should have been given to the utilities because after the research had been successfully completed and the projects moved into the practical stage the problems were of an engineering and not scientific nature. In the 1950s his had been a lone voice, but as the problems of the AGR became apparent towards the end of the 1960s, his views reached a wider audience willing to listen. Palmer and others saw that by shifting the civil side of nuclear research to the utilities, a lot of waste and unnecessary duplication could have been avoided. It was felt that the public corporation structure of the Authority was inappropriate as, unlike other public utilities and corporations, the AEA did not generate its own revenue (at least not on a large scale) and was therefore dependent upon its annual Parliamentary vote. The politicians were not technically literate enough, however, to appreciate fully the significance of the programmes they were voting upon, unlike the utilities; thus the AEA was able to avoid proper accountability in that, although it was dependent upon Parliament for money, the structural arrangements made it largely autonomous of both Parliamentary and ministerial scrutiny, whilst Parliament and Whitehall were dependent upon it for advice in all things nuclear. The AEA was able for much of its existence, therefore, to achieve the best of both worlds: Palmer and other MPs began to feel that the Authority had "got too big for its boots" and this was partly the reason for the decision of Ministers, acting upon the SCST advice, to hive off those parts of the Authority that were of a more commercial and applied nature. Palmer himself felt it would be more efficient to relate R and D to proven need.

For a period of over two decades the Authority was the repository of nuclear advice for the Government; indeed, it was synonymous with such advice. The AEA's raison d'etre became the promotion of their own reactor designs (see next chapter), and the enormous amount of resources in terms of money and manpower (and blind faith) invested in those designs made it politically difficult for the Government (of whatever political persuasion) to act in a contrary manner to AEA advice as this would have meant questioning the entire policy, a policy that had from its inception been linked to national prestige. A Government Chief Scientist in the late 1970s was able to argue that:

> "...this country should have a strong nuclear component is an old decision adhered to by a succession of governments. That there have been successes and failures is true, but in the various lobbyings that took place, that policy was never in doubt, either under Labour or the Conservatives, and my stint spans both. You can trace this policy back to the 1950s".

The basic structures of the policy were determined by Government decisions and their implementation determined by the technocrats, whose policy feedback determined future policy options: the extent to which they enjoyed political confidence is clear. This confidence was rudely shattered by the succession of British nuclear problems however and by the mid-1970s the Secretary of State for Energy was to tell the Chairman of the AEA that he had serious doubts about the quality of Authority advice (Williams, 1980, pp.243-244). As is shown below, part of the reason for the loss of political confidence was the failure for the Authority's advice when implemented as practice to provide the promised results; another factor in the loss of influence was the increase in the influence of other organisations, particularly the CEGB and some private corporations. This is also analysed in more detail below.

The result of the AEA's loss of a monopoly on advising Government from the mid-1970s was a slow increase in the number of other organisations permitted privileged access to the Whitehall machine. It was not a sudden change, but by the time the new Conservative Government issued its Green Paper arguing the point that to duplicate AEA advice in the DEn would be wasteful (1980), the Authority was already a fading star and having to compete to make its voice heard with the CEGB and business interests. Thus, although the formal structure of the AEA's role vis a vis the Government was essentially unchanged, the Whitehall ethos had changed and therefore the way in which that advice was received and acted upon was markedly different from that of a decade previously; the old deference had gone and the atmosphere of blind trust had dissipated. By the time the first Energy Secretary in Mrs. Thatcher's Administration, David Howell, took office the policy system was no longer rigid; indeed, there was a remarkably large number of basic decisions to make and these were made on political grounds, not expert advice.

By the time Howell reached the DEn, there was a plethora of advice from so many disparate sources that the Authority was just one of several repositories and one, moreover, that had proved lacking in the one quality above all others that politicians, prized, the ability to correctly foretell the future. It can be argued that the change in the influence of the AEA's scientists is a natural evolution of the role of nuclear power in society from a novel and esoteric source of energy to a mundane and vintage technology, such an evolution making the influence of the more commercially orientated the dominant influence on policy-making and therefore responsible for presenting the options to Government. As has been shown, although the internal structures of the AEA have changed, the constitutional relationship to Government has not; it is the informal relationship within the perceived norms operating in Whitehall that have changed. The influence of the AEA has declined due to the evolution of nuclear energy's commercial function; the engineers of the CEGB (see the next Chapter) have replaced the scientists of the AEA as the voice that whispers in the ear of Ministers.

THE AEA: AN OVERVIEW

The policy of the AEA is the product of the dialectic between political goals and technocratic advice tendered from a position of professional partisanism. The extent of technocratic subjectivity is explored in the following four chapters, but it can be strongly argued that policy proposals are the result of professional objectives framed within the terms outlined by stated political goals. This is a hypothesis that suggests itself from the analysis of the succeeding chapters. The goals of successive governments obsessed with strategic independence in energy supplies, nuclear nationalism and exports, have permeated down into the fabric of decision-making. Even the scientists of Harwell were to shift towards the prevailing quest for export-led profits, the commercialisation of the Production Group spreading throughout the structure. Whilst in the days of Cockcroft the scientists may have been commercial virgins, their state of innocence was rudely intruded upon in the 1970s when their Director, Walter Marshall, injected into their considerations the cold logic of 'economic reality', a logic that found favour with the administration of Mrs. Thatcher and saw him ennobled as Lord Marshall, the Chairman of the CEGB. In 1974 Marshall wrote:

> "...the necessity to earn money from industry has high intrinsic merit by itself. It makes all the staff of (Harwell) more realistic, it increases their general knowledge of technology and forces them to be more balanced in the judgements". (Atom 210, April, 1974, pp.80-81).

It can be seen that the federal structure of the Authority that originally reflected the differing world views of the engineers and scientists, and the differing world views moreover of the original leaders of the Groups, evolved into a more centralised and consensual structure. Even as the engineers were leaving the AEA for the commercial world of public corporations, the Authority began to succumb to the persuasiveness of their arguments.

That the desire for commercial success is subservient to the broader professional goal of occupational autonomy is a recurrent them of succeeding chapters and is also suggested by further remarks of Marshall, when he stipulated that the aims of the technocrats included the primary aims of the maintenance of their research teams and the autonomy of those teams. He argued:

> "...when a group of talented scientists and engineers have been brought together into a national laboratory and they have partially fulfilled their original mission, it seems to me unimaginative and wrong simply to disperse the ability which has thus been gathered together. In short... they should be encouraged to change their orientation from time to time". (ibid)

In the Groups concerned with production and with fusion research this reorientation included changing the goals not only of the teams but also the organisational objectives or those of the Government. Marshall recognised the extent to which the reorientation involved a break with the accepted professional ethos of science to those engaged upon research at Harwell, when he noted that in his personal experience "changing the programme and orientation of a large laboratory" is difficult "because scientists are very reluctant to abandon their own pet line of research" (ibid). A former Chief Scientist argues that:

> "scientists are the biggest bloody snobs there are in that they are quite content to be said 'no' to by someone who has won some distinction in science, even if it is in a field that is a long way from their own expertise. But they are not content to

accept such guidance from a non-scientist. You can ascribe this attitude to snobbery, but there is more to it than that".

The point being that scientists, like others claiming professional status, prefer to be self-regulating (see previous chapter), the ability to change their own goals and workplace projects being a cherished part of their occupational autonomy. Thus the managers leading the teams and seeking to interpret the policies of the politicians have traditionally been technical experts themselves; other arrangements would not have facilitated the smooth functioning of the research teams. Even Allen, who has a non-scientific background, converted to the norms of the Authority, otherwise it is probable that his appointment would have led to a crisis of confidence even greater than that which already afflicts a body subject to some formidable criticisms in recent years. Perhaps the change that has occurred in the Authority can be summarised by two quotes from Marshall:

"When we began the programme (at Harwell) we had very little commercial experience ourselves and while that has been a handicap, it has not proved to be disastrous" (op.cit.),

presumably because he now feels that:

"...a good basic scientist will be a good business entrepreneur if he can be persuaded to make the transition and is given the right environment and encouragement to do so" (op cit).

It has been the function of the senior managers in the Authority almost from its inception to seek to persuade the engineers and scientists they supervise that such a conversion is both advisable in the interests of occupational development and also compatible with the professional aims and ethos of those involved.

Within the structures the technical experts at all levels sought to influence policy and to have their goals adopted as those of the Authority. The question of policy organisation is a vexed one and has been the subject of some complex analyses (Vickers, 1965, 1980; Ashford, 1981; Burch and Wood, 1983); within the Authority, as elsewhere it is a mixture of action and reaction. That is, the groups react to external stimulus, for example Government cuts in funding and they also act to secure goals through presentation of proposals. A senior manager with the Authority argues that policies emerge in a variety of ways, through scientific discovery, through the needs of client groups such as the utilities or the Government, or through the results of market research. Policies are aggregated through discussions at committees of the AEA and through joint committees with CEGB and DEn; most therefore originate within the discussions of middle and senior management acting upon the technical recommendations that emerge from within their research teams, or reacting to external pressures. Of the former:

"...when someone comes up with a bright idea he has to sell it to the management and because the management are of a mostly scientific background themselves they are able to lend a critical ear to the proposals. If they do not come up to standard then they are quickly dispelled. It must be remembered that the AEA has only a limited fund and therefore if a new idea is to be funded then money must be taken from existing projects and these will have tenacious defenders".

This description rather simplifies a complex process whereby policies emerge from a morass of technical and political input, the amount of each varying according to the subject being debated. Often the non-decisions are as important as the decisions, policy-makers subconsciously adding or subtracting considerations that are the product of internalised norms, norms themselves that are produced by their political, organisational and professional socialisation (there is a copious literature on this: see the previous chapter for a discussion). The Authority's manager pointed out that:

"...decisions are made along political, technical and economic lines. The DEn receives a lot of information from the scientists of the AEA. The civil service aggregates, consolidates and screens this information before it goes to the politicians, if at all, so that ideas and policies are aggregated all along the line. Because of this the line management system of the AEA is important. Engineers and scientists as managers are very important because of their ability to turn a critical ear upon their subordinates."

Policy proposals, then, emerge rather like spirits from a still; they are the result of a distillation of a variety of factors, not least of which for those generated internally is the line management structure which subjects the proposals of subordinates to a lengthy and often inflexible process of scrutiny, with at least one middle manager complaining that the rigidities in the system rendered it inflexible and inefficient. A former Production Group manager argued that policies were imposed from the top and when proposals originated at the lower reaches they met with tremendous opposition from the vested interests:

"When the scientists took over the industrial chemistry group I reported to a superintendent, he reported to a senior superintendent and he reported to a deputy director. All of these I'd known for some time, but those three were scientists and interested in making a lot of scientific kudos out of spending a lot of research effort doing things that to my mind were unnecessary and time-consuming. We therefore got into a conflict situation over this. The Deputy Director at first supported me but then he called me into his office and said "You may well be right, but I am going to have to support your supervisors because they are superior to you"."

An example of the inefficiency which he argues was caused by the rigidities of the system occurred when it was decided to give the job for producing plutonium ceramics for the Fast reactor, not to the Northern factories but to Aldermaston instead. He said that the people in the North:

"...like Sir John Hill and Franklin called me into their office one day and said they were so concerned at this job going to Aldermaston in view of Aldermaston's appalling safety record, that they wanted me to go down and run the plant".

The manager believed the decisions to transfer this process to Aldermaston had been taken by Sir William Cook as the Aldermaston teams were running out of work and Cook being a former Deputy Director there had sought to provide a new line of work to keep the teams together. Furthermore, such an esprit de corps functioned throughout the federal structures of the Authority, as when the manager wrote a report showing how the Authority as a whole could save money by scrapping a proposed extension at Springfields on the grounds that existing facilities at Dounraey were adequate. He argues:

"When I put this to my boss he was furious. He said that Dounraey was not in the Production Group and my responsibility was to the Production Group, he did not want to know about the Dounraey equipment... We were on the fourth floor and I would write a paper which would go to the Production Group bosses on the fifth floor and I would never meet anyone from the fifth floor. But I could only pass my report up through my head of section and as a result my papers were severely changed before they reached the fifth floor. The report on Springfields did not get to the fifth floor for a long time. It was a year before the project was cancelled, but by that time they had spent £1.2m and I had to use unofficial sources to get my report noticed."

Whilst the details of these examples may be open to question, the sentiments are not so easily dismissed; that is, the rigidities inherent to the structures of the Authority for much of its existence severely influenced the policy options that emerged or were put before senior management, and thence to the policy arbiters in Whitehall. The system was such that effective options were precluded from discussion because people believed they were not options that the senior managers wished to be discussed, or because they were perceived to conflict with occupational/organisational goals. The sort of screening that inevitably occurred is consistent with the work of organisation theorists such as Downs (1967, pp.49-74) and the hypotheses of the previous two chapters, in that although organisation theory explains much of the behaviour involved in the structuring and initiation of policies through the communications networks, a full understanding of the motivations involved only comes with the inclusion of professionalising ideologies into the explanations.

In conclusion, it can be argued that the role of the Authority and the influence of its experts within the policy system has been dynamic; initially of great import but then tending, like the AEA itself, to entropy. The evolution of the system as a whole is partly responsible for this in that the AEA ceased to be the sole repository of nuclear expertise within the U.K. and those organisations, both public and private, which began to rival the Authority for the ear of the Government tended to be more in tune with the commercial and economically expansionist goals of the politicians; they also had the engineering and commercial expertise to give better advice to the politicians about how to realise those goals, the scientists of the AEA being late converts to the commercial ethos that prevails outside of the hallowed portals of universities and research establishments. Those groups within the Authority that were able to adapt themselves to the commercial applications of nuclear power, namely the Production Group teams, sought to have themselves hived out of the Authority in order to ensure their continuing occupational independence and prosperity, reasoning correctly that they would have greater independence to pursue their professional goals if they were free of the restraints imposed by the Authority, a body that had evolved for a purpose quite different from that which many of its constituent parts had come to perform. The role of the AEA experts in tending policy options for the nuclear programmes is one that began with them exercising a monopoly and has reached a stage where their influence, although still great, is now outweighed by the other players in the system; they are in a sense a Greece to the CEGB's Rome.

It is the rise of the CEGB's influence within the nuclear energy policy-system that provides the focus of the next chapter. It largely mirrors the events portrayed here, the Board being a professionalised organisation dominated by its technical experts and anxious to convince the policy-arbiters of its goals. In reviewing it after a discussion of the AEA a better perspective is obtained upon the policy-making role of the technocrats, the two organisations being very different but interlocked within the policy process. The Board's managers learnt their political lessons from viewing the initial dominance of the Authority and it is the implementation of those lessons that is analysed next.

5 The professional structures II

As the concomitant to Chapter 4, this chapter is intended to probe the professional structures of the CEGB, structures that increased their political influence as that of the AEA dwindled. But in addition to simply charting the role and position of the technical experts, the latter sections also attempt to explain the complex relationship between the Board and the Authority, and the Board and politicians and industrialists. The conclusions of this chapter, therefore, are germane to the analysis of the previous chapter too; indeed they are informed by it. The chapter's first task, a description of the CEGB, leads onto a review of the role of engineers within the hierarchies of the Board before assessing their contribution to policy making. The wider concerns of the relations of the CEGB to Whitehall and private industry then give way to a consideration of AEA/CEGB relations and the contrasting style of each organisation, before concluding with an attempt to put the role of the CEGB into some perspective within the nuclear policy system and review the influence of technocrats (in both the AEA and CEGB) in policy making.

In repeating the format of Chapter 4 in its initial sections, that is, by first outlining the structure and functions of the organisation being researched, the intention is to explain the framework within which the technocrats are situated in order to understand the context from which they operate. In the case of the Board the dominant technocrats are engineers, and the reasons for this are discussed with reference to the review of engineering in Chapter 3. Once these preliminaries are complete the way is clear for a more detailed analysis of the actual (as contrasted with the formal) role of technocrat/political relations left off at the end of the last chapter, in particular there are some arguments as to why the CEGB has supplanted the AEA as the main nuclear adviser to the Government. Some of these reasons involve the role of private industry in power station construction and a section examines these, before the chapter concludes with observations on the role of experts in the policy-process: a section that prepares the ground for the in-depth research of reactor decisions reported in Chapters 6 and 7.

THE CENTRAL ELECTRICITY GENERATING BOARD

The CEGB is the largest organisation within the electricity supply industry; the other actors are the two small Scottish and one Northern Ireland Board (of which only the South of Scotland Electricity Board possesses a nuclear component) and the English regional boards, to which the CEGB sells its product for resale to the consumer. Some of the largest consumers, such as automobile plants do however purchase some of their electricity direct from the CEGB, and the SSEB, NSHEB and Northern Ireland Board sell direct to the consumers. The ESI is a nationalised industry and the CEGB came into existence on 1st January 1958, succeeding the Central Electricity Authority under the terms of the 1957 Electricity Act. It was born into controversy, part of the reason for the reorganisation being, according to Hannah, "a deeply-felt desire within the industry" to free itself from the central control of Sir Henry Self and Sir Walter Citrine, the two men who had led the nationalised industry since it had been taken into public ownership by the Government of Clement Attlee (Hannah, 1982, p.184).

The first Chairman of the CEGB was Sir Christopher Hinton who moved from the Production Group of the AEA, a move symbolising the commitment of the utilities to nuclear energy, or at least the commitment that the Government had for the utilities. In a sense the CEGB was not so much a successor to Citrine's CEA but a continuation of it. Citrine had taken over a system in the late 1940s where the profit motive was already small and with Self (a civil servant) he began to put into practice the corporatist ideals of his Labour and trade-union background. He set up the centralised machinery which survives to the present in a modified form; the structures of the industry reflected the desire for participatory decision-making combined with a centralised system designed to maximise independence from political control. It was run along collective technical lines (ibid. pp.1-16) and reflected the Morrisonian and Fabian principles of Citrine's TUC background in the 1930s. The early years of Citrine's control saw a struggle between Self, Citrine and the engineers of the industry, led by men such as Hacking and Pask (ibid. pp.96-109) who sought to frustrate attempts at administrative control, particularly in areas of technical decision-making. Citrine and Self won, and in 1950 set up a contracts department, attempting to impose some commercial efficiency upon the technically-minded, but commercially ignorant engineers.

With the appointment of Stanley (later Sir Stanley) Brown as chief design engineer in 1954, the tide was turned. Brown instigated a shift in engineering thinking and practice that became known as the 'Brown Revolution', ordering generating sets of increasing size and complexity, forcing through economies of scale in all areas of generation and ensuring the consolidation, expansion and rationalisation of the industry. Under the leadership of Citrine the electricity supply industry doubled its capacity in only ten years, interruptions in supply became rare. Other reforms included the imposition of a trade union closed shop agreement, which greatly facilitated negotiations between the management and workforce, leading to superb industrial relations throughout most of the ESI's post-war history.

A second reform, dealt with in greater detail below, was the formative attempt to adopt a nuclear element into the generating plant of the industry. From 1954 Whitehall became increasingly committed to the idea of a nuclear component and the working party set up under Burke Trend led to the White Paper of 1955 announcing the first programme of 1700 mw of nuclear power by 1965. A nuclear power branch was established under Duckworth at the British Electricity Authority (the precursor to the CEA) and the industry agreed to the Cabinet's proposals somewhat reluctantly (ibid. pp.168-175). Engineers began to be sent to Harwell and Risley where the scientists of the AEA trained them in the use of nuclear technology for civil power programmes.

Before proceeding to discuss the role of nuclear power in the generating capacity of the CEGB it is worth remembering that this source of energy is only a small, albeit increasing, component in the total capacity of the ESI. In 1980 the Board had a power system comprising 132 power stations with a net capacity of 57,029MW and a national grid network of 14,659 km. of main transmission lines (9,109.1 miles) which was, in the Board's words, "the largest fully-interconnected system under unified control in the Western World" (CEGB, Electricity Supply in the United Kingdom p.16). In 1980 the Board had under construction a further 13,430 MW of plant including 3,840 MW of nuclear capacity, most of the new plant (excluding the pumped storage stations) consisting of several 660MW generating sets designed, inherited at nationalisation and constructed during the 1950s, of less than 550 MW, which no longer provided the sort of efficiencies demanded. 77 per cent of the fuel consumed for generation in 1980 was coal and the total fossil fuel burned in that year accounted for 88 per cent of the Board's output. Nuclear power accounted for about 11 per cent of the Board's output with the rest coming from gas (ibid. pp.17-20). The Board predicted that the nuclear component would rise to nearly 20 per cent before the end of the decade. By 1986 the Board had only 79 power stations and although several of the AGRs were nearing completion the nuclear component remained similar to that of the start of the decade, having increased only slightly. With the planned retirement of the Magnox, the Board was anxious to order up to 10GW of plant before the end of the 1990s (CEGB, 1986). These figures contrast with those of 1948 in that there was no nuclear component, there was a capacity of 10,363 MW, a grid about half the size and nearly all the electricity was produced from coal in old, inefficient 30 MW and 60 MW sets (situated throughout over 600 separate entities inherited from the pre-nationalisation structure). Interruptions were frequent, at any one time up to 20 per cent of consumers could expect to be disconnected as there was a severe dislocation in supply and demand throughout the country in addition to an inadequacy of total supply.

The situation was such that it had electoral repercussions and the CEGB has been obsessed with continuity of supply throughout its history: indeed, it is enshrined in the 1957 Act that the Board must seek to ensure continual supply at the lowest possible price. Thus the CEGB states that its objective:

> "...is to provide a secure and economic supply of electricity and it therefore manages its assets and undertakes new investment both to provide sufficient capacity to meet demand and to supply electricity to the consumer at the lowest cost consistent with security." (1982, Evidence to Sizewell 'B' Inquiry, vol.1, p.1).

To this end, the Board has interpreted the lessons of the Suez crisis, the Yom Kippur war, the national Union of Miners' strikes and rising coal and oil prices to mean that it can best meet its brief by substantially increasing the nuclear component in its generating plant at the expense of the fossil fuel component. The Board seeks a strategic independence and flexibility of supply; it therefore feels that nuclear power which is not dependent upon miners in Britain or Arab oil cartels is the optimum fuel for generating electricity, particularly should the fast breeder reactor with its superb efficiencies come on stream as a commercial proposition.

The raison d'etre for the Board then was plain and permeated throughout the organisation: the dedication to the cause of a vast expansion in the 1960s and 1970s led to the growth of a close spirit of shared commitment within the ranks of the engineers. A former senior engineer and Board member recalls:

> "In the 1960s and 1970s there was a tremendous esprit de corps within the CEGB; it was probably unique. I used to sometimes go into the office on a Saturday afternoon or Sunday morning and there would be the drawing office boys beavering away in what was completely voluntary work: they weren't paid at all after time but they felt such a commitment to the job and were so dedicated that if something had to be completed quickly they would work all hours to do it. This spirit permeated throughout the organisation. If there was a storm and power lines came down everyone would swing into action very quickly, everyone was, and still is, dedicated to keeping a continual supply of power flowing."

Thus as the CEGB grew rapidly in the first decade of its existence after 1958, the technical experts felt an attachment to it that was stronger than that of mere employees to an employer; it was an attachment that grew from the role of the Board as an engineering organisation par excellence. As is shown below, the Board was run by its engineers as an engineering organisation serving engineering goals; it became a visible manifestation of the profession, or at least those parts of the profession concerned with electricity generation, motivated by something 'higher' than the profit motive of private industry. This is explored in more detail when the role of engineers within the CEGB is examined.

Engineers within the CEGB

The structure of the CEGB at national and at regional level, although modified, has not experienced drastic changes over the last twenty-five years. It is a standard line management structure with formally-drawn spans of control all the way down the pyramidal hierarchies encompassed by the organisation, the peak being the Board itself. Ince is probably overstating things when he refers to the CEGB as 'a labyrinth' (1980, pp.23-27), but it is certainly complex - necessarily so in that although the employment peak of over 80,000 (Annual Report, CEGB 1966-7) has been reduced to about 56,000 (Annual Report, CEGB 1981-2), the Board has responsibility for a formidable number of people. Its main functions are the planning, building and running of transmission lines and power stations, as is clear from the points made above, and to this end it employs a large number of engineering experts, as well as cohorts of other professionals such as physicists, economists, statisticians and accountants. It is the engineers, however, who are the senior experts within the Board. The staff and line structure enshrines and augments their dominance.

Just how dominant the engineers have been is illustrated by the remarks of serving and former Board members. One of them recalled that:

> "All the key people in the CEGB were always engineers. It wasn't that people thought themselves engineers or other-wise, it was automatic: people were engineers. It was like breathing, they were engineers full-stop. All you then concerned yourself with was the ability of that man to fill that particular role. I remember one man who came from the AEA and his particular grouse in life was that he was paid as an engineer; he maintained that he was a scientist and he had come to us in a very senior research role. But we took the view that as far as we were concerned he was in an engineering institution and therefore he was paid as an engineer; end of message."

The senior managers in the CEGB were therefore fiercely resistant to any attempt to dilute the engineering content of the upper reaches of the hierarchy, a situation that had led in the days of the BEA and CEA to some bitter clashes with Self and Citrine as the engineers resented those from outside of their technical sphere seeking to intrude. The above quote is also interesting for the inkling it gives of the relationship between AEA scientists and the Boards' engineers, a subject analysed in more detail below. A former high-ranking CEGB finance executive also noted the engineering dominance:

"It's changed, but until fairly recently it was better for your career prospects within the CEGB if you were an engineer. I've been in this industry for forty years and there was a time when the only chap who was right was the engineer - he knew all the answers to everything. It took a long time to get them to be cost-conscious. In the early days when I tried to get them to introduce budgetary control, one of the old engineers said "We don't need any of this nonsense. I'll look at the results at the end of the year and that's all I'll need". I had a tremendous battle to get them to see that they needed proper cost estimates. That has changed now and I think they see that the financial man and other experts have got a proper role to play in decision-making."

The ethos of the CEGB, then, is one of dominance by the engineers, perhaps even aggressively so; these are design engineers and the multitude of disciplines involved in the construction of power stations and transmission of electricity. The situation was encapsulated by a former Chairman of the Board, when he argued that the CEGB as a technical organisation depends upon its engineers; to a large extent it is its engineers. No policy decision can be taken without a tremendous input from them and it certainly cannot be implemented without them. They are responsible for formulating its policy and this is then subject to other pressures such as commercial and political interests.

Throughout its history there was no point in the hierarchy of the CEGB at which it was possible to distinguish between engineering and management positions, at all levels in the organisation you could get an engineering answer. One irony connected with this engineering hegemony is that the CEGB also has massive scientific research facilities at Barnwood, Marchwood and elsewhere; indeed its scientific resources dwarf those of most British private corporations and are equal to those of most Government-run institutions. In all probability the CEGB employed more scientists than the AEA, but it was the applied and practical nature of the Board's mission that dictated that they remain subservient to the requirements of the engineers. The allegiance of the engineers within the CEGB is such that it confirms the dual allegiance hypothesis of the last chapter and of Dunleavy (1980b), in that the vigorous attachment to their professional calling is displayed in tandem with an equally strong attachment to the Board, as was noted above. A member of the Board argued that whilst "one is always an engineer" that does eventually tend to become compromised by two things:

1. the need to become a general manager, even though you remain an engineer at heart, and
2. in this organisation the need to become a CEGB man, not to be interpreted as a brainwashed man, but to be wholly aware of the organisation, the way the organisation works, what its strengths and weaknesses are.

He argues that therefore:

> "...people inhabit two worlds; they retain a lot of their engineering professionalism and lot of their engineering expertise (that's vitally essential, we wouldn't half be in trouble if they didn't), but on the other hand they have to assimilate the sort of ambience of the organisation... and their engineering has to be applied and thought through within the context of the organisation. That in itself is within the context in which the organisation works.

Clearly the links with organisation theory mentioned above are apparent here also, and the importance of "non-decisions", as in the work of Bachrach and Baratz (1962). These themes are returned to below, the essential point here being that the dual allegiance of the technical experts, subsumed as they are within bureaucratic hierarchies, is recognised and encouraged by those organisations primarily concerned with a technical function. The CEGB is an example of this type of organisation and it is one moreover where the hierarchy is colonised and directed by the technical experts; indeed they use the structures to support their own position, organisational and professional goals being part of a continuing dialectic.

Engineers as Policy-Makers

As with BNFL (see Chapter 8) it can be strongly argued that the ESI is a professionalised industry and the policy options it initiates and pursues reflect this. Commentators such as Sweet (1982), Pearson (1981) and Williams (1980) have accused the CEGB's policy-making processes of a host of anti-social traits, Sweet arguing that they have deliberately misled on economic grounds (op.cit. p.33), Williams noting the lack of accountability and participation (op.cit. pp.312-331), although he accused the whole policy system of this. Pearson pointed out that the CEGB simply did not believe in public participation in policy-making and that in 1973 Hawkins said that issues of nuclear policy should be left to the technically qualified (op.cit. pp.88-93). Such criticisms seem well founded and indeed these traits are to be expected in activities dominated by experts, as is clear from studies in other areas, such as Wilding's analysis of the health professions (1982), Saunders's work on the Water Boards (1983), and Dunleavy's work on local government (1980b). One of the primary goals of professional occupations in achieving work autonomy is to reduce non-expert input into policy-making and then to seek to have policies made on technical, non-political, criteria.

The CEGB is aware of the reputation it has gained and there is strong evidence (see the evidence to the Sizewell Inquiry, and Kemp, 1985) that the Board is attempting to move away from its image of monolithic autocracy. Certainly at Board level there is a desire to acquire a more open image, a former Board member argued:

> "CEGB tends to have a history of being rather arrogant and self-sufficient, and indeed we are self-sufficient and technically very strong...until really up to the time that we began to consider preparing proofs for the Sizewell Inquiry we

were really too insular. But over the last three or four years we've realised that, and we consult people like the London Business School and the Cambridge Economists."

The engineering input into policy, however, remains over-whelming. Yet the way in which policy emerges within the CEGB is not as formal or monolithic as staff and line diagrams and the criticisms of outside observers would suggest.

From the time of the 1955 White Paper which established nuclear power as a civil option for the first time in Britain, the ESI began to send young engineers to the AEA for training and they, along with their successors, had an important input into policy discussions on nuclear matters. In a sense, the Board itself, rather like the politicians and civil servants of Whitehall, is a policy arbiter, choosing between the options its engineers put before it, before honing them into hard proposals for presentation to the rest of the policy community. As with the AEA, policy emerges as a result of action and reaction; action in order to achieve perceived needs or solve perceived problems, such as inadequate electricity supply; reaction to Government decisions, such as Varley's 1974 decision to foist the SGHWR upon a unwilling CEGB (see next chapter). As a broad generalisation, it is almost impossible to pinpoint when particular decisions were taken because, as one former Board member argued , "they evolve" within the framework of society generally and the particular policy communities. He argued that with the nuclear decisions:

> "...somebody talks to somebody and then an idea forms and they you have a word in the club and the whole thing gradually creates an atmosphere which may take months, it may take years, to evolve."

The genesis of ideas that turn into policy initiatives emerge as the result of professional interactions with regard to the practical potential of technical innovation; that is, experts ask each other to what use their knowledge can be put. This is, however, just one method by which policy emerges within the engineering ranks and it is for the most part a rather insignificant source after the initial discoveries in a particular field have been made. For example, within the nuclear field, this type of innovatory discussion would have occurred in the 1940s and 1950s.

Nearly all the parts were decided by and evolved by definition out of the military programme. The major innovator was Hinton. Once these strategic options were decided upon then policy making became a continuous process and it is here that the engineers have a major input on a continuous basis, the strategic initiatives being more openly political, if not in content then in potential.

It is within the region of daily problem-solving that policy options tend to emerge; in a sense much of the Lindblom thesis regarding disjointed incrementalism can be glimpsed in the workaday process of a large organisation as it seeks ad hoc and satisfying solutions to problems (Lindblom, 1968;1977). Yet the same critique as can be applied to incrementalism per se can be used against this view, namely that it is a behaviourist interpretation of complex relationships, failing to understand individual and group motivations and being unable to provide a satisfactory explanation of the fundamental decisions, or even pointing out where the distinction is between incremental and non-incremental change. Essentially, policy-making is a mixture of action, reaction, strategic intervention and simple muddling through. In all of these processes the engineers provide the technical basis for decisions. The formal structures are the legal outline within which communications flow; yet there has always been an informal network that greatly increased (and increases) the potential for young or junior technical experts to influence policy, sometimes quite considerably. In addition to the informal network there are ad hoc formal committees which set up working parties and sub-committees to review aspects of problems or policy initiatives; these cut across the formal lines of the hierarchy or even across organisations themselves, a recent influential example being the structures established that eventually led to the Thermal Reactor Strategy Document (CEGB 1981) that outlined the Board's and the Government's proposals for introducing a programme of PWRs into Britain (see next chapter).

All senior Board members who have commented upon the role of engineers in shaping policy have remarked upon how they had a direct input into policy at a very early stage in their careers, sometimes in their mid-twenties. One ex-Board member recalled that:

"I felt that I had an effect upon policy-making long before I reached Board level. The way I wrote reports, the research I did, the people I reported to, all had an effect. Of course, until I reached Board level I had to be very careful about the way I presented papers and opinions to my superiors, but I and all senior researchers and engineers affect policy, of course they do, that is the purpose of their advice. When I came to the CEGB I was ....'s personal assistant and that gave me influence."

Again, this reminds us that advice, proposals and technical information are constantly couched in terms that the vendors feel will appeal to those in authority. Proposals and advice that it is felt will stand little chance of consideration or which lacks a sufficient political appeal is either screened out or distorted. As is clear from a later section in this chapter, such methods were often followed one ex-manager arguing that although the CEGB did "a massive amount of research" it "didn't always publish all the data" and that sometimes when they did publish it they "put a different emphasis on it". That being, as he said "politics", in that

"...when we were asked to provide information we always tried to do it so that we got people to do things we wanted, we always tried to win decisions.

Thus, the training given in policy presentation from an early stage in their career can stand to profit the engineer-manager later on when dealing with the political system.

In a sense the Board operated a form of democratic centrism in that at an early stage of proposals alternative advice was openly sought from within the ranks. The process of discussion involving the alternatives tended to produce unity, one ex-Chairman even arguing that those who did not agree with Board decisions after the discussion process would be offered the chance to "go off and do some research to prove or disprove their theories" (ibid). In practice, mavericks have not always enjoyed such luxury, Ross Hesketh, a senior research physicist, being sacked in June 1983 for consistently and publicly stating that he believed waste fuel from Britain's reactors had been used in the American weapons programme (New Statesman, 17/6/83). Although this was not strictly a policy disagreement, it is indicative of what rebels could expect should they fail to take advantage of opportunities to recant; Ince perhaps overstated the situation however when he argued that:

"...the corporate structure of the CEGB is unanimous that the cheapest and best way of achieving an efficient electricity supply system is nuclear...people who disagree do not last long in the CEGB." (1980, p.23)

It appears that alternative viewpoints have traditionally been sought as to how the Board should proceed; these have been followed by sometimes heated discussion until agreement has been reached and then solidarity has been expected. Since the Board cannot be expected to embrace frequent changes of policy, by the very nature of its function and the content of its work, there has evolved something of an inertia in policy terms, the CEGB gaining a reputation for inflexibility.

One problem with the informal communication networks is just how a disgruntled expert promotes his ideas and pushes them through to his superiors if he feels that the formal staff and line hierarchies are blocking him or stifling what he feels to be important innovation. One way is through using the professional institutions and associations as a conduit of communication through which ideas and proposals flow. In this sense the learned society function of the professional institutions are important both as a channel to place viable policy options on the agenda, and perhaps as a safety valve to allow people to 'let off steam'. One CEGB Board member argued that, in the view of senior managers, the professional institutions:

"...become the centres of technical excellence which are drawn upon. They also become an outlet for a chap who has particular views on the things he feels ought to be done and he finds that his immediate management are a bit of a block (and none of us is perfect), so he is able to bypass his management."

It is clear that over the last decade and particularly since 1983 the CEGB has sought to instigate a more 'permeable management training and management attitude' (ibid). The new regime under Lord Marshall is symptomatic of this in that there appears to be a flexible style of middle and upper management, something akin to a 'leapfrog' style of informal contact where a subordinate feels free to bypass his immediate superiors at any stage of the hierarchy in order to deal directly with higher management:

> "Take my own case as head of this Department. The man under me, say one of my branch heads, knew more about his responsibilities than I did. He knew less about the political implications about what he was doing and the ramifications for the Board's general wellbeing about some of the things that he might be postulating should be done. If a member of the Executive sought advice on a particular subject I found it was much better to let the chap who was working for me and had a knowledge of this subject to go directly to the member of the executive. If he came to me and then I relayed it the chances are that the member of the Executive would ask a detailed question which I couldn't actually answer...and if I want to learn something as likely as not I will go to the man down the line plus one or even two," argued one senior executive.

This then is the way in which engineers feel that they came to be an influence upon policy-making even when in comparatively lowly positions. The informal networks so necessary for a free flow of communication, the raw material of policy-making, gave them an access to the higher reaches of the management structure, providing of course that the information they supplied and the recommendations they tendered were in accord with the general policy of the Board, that in turn being shaped by the rougher elements of the political environment in which it existed. The early career of one ex-Board member who was a pioneer in the civil nuclear field is an example when along with five other young engineers, all in their late twenties and early thirties, they were sent off first to Harwell and then to Risley as the pioneers within the ESI for training as nuclear engineers. He described how these six became the key people for the CEGB's nuclear programmes, taking on overall planning, operational and reactor engineering functions:

> " What the Board effectively said was 'We don't know what the bloody hell it's all about, but don't land us in the cart, go away and do it'. The exception of course was Hinton" .

In the early years a tremendous degree of freedom was therefore granted to these pioneers; in the late 1950s they wrote the safety and design rules as they went along. For non-political policy areas this type of freedom for the technical expertise is the norm and they expect it as their professional due:

> "As long as there's a dynamic demand by society for an increase of the product, an increased technology, society has to respond by putting into office the people who have the necessary professionalism to do the job, it has no option...however much it may distrust them," argued one CEGB manager.

As becomes clear below, the distinction that the experts tend to make between the technical and political arenas is often a false one. Their belief that they should have the freedom to proceed as their technical expertise dictates is itself a political position fraught with subjective and partisan implications, as writers as diverse as Habermas (1971, 1976), Wynne (1982), Dickson (1974) and Price (1965) have attempted to show. Theirs' is an essentially flawed view of democracy; it supposes that once a decision has been made about general policy direction such as whether to proceed with nuclear power or not, then the technical details should be left to the experts and political considerations should wither away. This is a reflection perhaps of the style of decision-making within the CEGB itself, it certainly led to considerable frustration being expressed by CEGB technocrats whenever politicians had the 'audacity' to intervene in 'technical' matters, notwithstanding the political import of those matters (see below and Williams, 1980, p.245, and various interviews).

It can be argued that the deciding input into the policy deliberations of the Board is that of the engineers and at Board level the technical proposals of the engineers are assessed for their political implications, a factor noted by one Board member who said:

"The Board as whole forms the policy...you've got make sure that you are not being blinded by science, you've got to make sure that you are asking the right questions".

This led him to conclude that although the engineering input is the significant input, the 'policy of the Board is not the policy of its engineers', a view that could perhaps be termed the wishful thinking of a non-engineer, especially in the light of those career experiences quoted above. Indeed, his argument that, at Board level 'one asks the right sort of questions' such as the possible costs, safety aspects, economic impact on other parts of the economy, etcetera, actually confirms that the Board's concern is technical, plus the political or the potentially political implications. The impact of the political implications is then weighed and the technocrats seek ways to minimise it in order to retain the clever fiction of non-political policy-making that supports their claim for technical precedence in the decisions taken. An example of this examined below is the CEGB's horror at the decision by David Howell to announce that the Government of which he was a member intended to build fifteen PWRs, an announcement one CEGB executive described as bringing 'all hell down upon the CEGB and the nuclear industry'. Within the CEGB the hierarchies of the bureaucracy contain the technical experts; indeed they consist of the engineers plus a sprinkling of administrators, accountants and scientists. The prevailing ethos it that of the engineers, but their allegiance reflects the norm of the occupations in that it is both to their profession and to the organisation that employs them, the goals of each acting and interacting upon each other. Thus whilst it could be said that the policies of the CEGB are its own and not those of the engineers it employs, it must also be said that those engineers provide the information upon which decisions are taken; theirs are the dominant ethos, the methods of operating and the methods of research.

In short, although the parameters of policy-making are determined by the political environment in which the CEGB operates, its policies are those of the engineers it employs most able to interpret the political situation in terms easily transformed into technical options. That is, if they can present their advice in such a way that it accords with the overall goals of the policy arbiters, then that advice will be followed and the engineers concerned will be given considerable autonomy in its implementation. As one Board member argued:

> "If I seek advice from one of my staff on what might be thought of as a straightforward engineering matter, say the conversion of Grain from oil to coal firing, what I don't expect is a piece of paper with economic evaluations which start with the capital cost of the conversion, and go through the evaluations of the future of the energy economy and the load factor of grain etc. I would expect to get these but I would also expect something about the impact on the coal industry, the impact on our own oil burden, the effects of a miner's strike if we ran out of coal. So we expect from our engineers a different approach to what are basically technical problems according to where they are in the organisation. As they progress up the hierarchy they must educate themselves politically".

In essence, in seeking to avoid politically contentious decisions or at least presenting them in a way which suggests that they are technical options unworthy of a politician's time, the technocrats are behaving in a necessarily politically astute manner. The prize of professional and organisational autonomy is won only by those best equipped to play the political system, something that requires more than just a skill in engineering.

The Board and Policy-Making

Before analysing the relationship between the CEGB and Whitehall in more detail, a brief review of the Board's evolving view of nuclear energy is useful in that it aids an understanding that is also relevant for the analysis of the next chapter with regard to the reactor decisions. One of the interesting characteristics of the Board's decision to change from a grudging acceptance of a nuclear component in its generating capacity to an enthusiastic embrace of nuclear power, is the people who changed their views. This is particularly important within the context of the Board, as the most influential of these became Chairmen; they include Hinton (1957-64), Brown (1965-72) and Hawkins (1972-77).

The analysis above has explored the role of engineers *per se* in the policy-making process of the CEGB; the role of the Board and of the Chairman is one that is changeable, although the centralised decision-making machinery has ensured that once the Chairman had made a decision he was able, with the backing of his Board, to impose it. Hannah explains the process whereby the Government in the mid-1950s 'encouraged' the ESI to accept the nuclear programme for strategic reasons which included a fear that coal production would be outpaced by demand, that oil prices would rise dramatically, a fear compounded by the 1956 Suez crisis that saw the vulnerability of Middle East oil exposed, and by the desire to flaunt a technical expertise at the rest of the world (1982, pp.168-175). Initially although Citrine accepted the proposals as a political inevitability and encouraged the engineers to do likewise, the technocrats within the BEA and CEA were opposed to the proposals. In this they were led by Brown who wished to concentrate on consolidating the impressive strides he was achieving in improving the efficiencies and capabilities of the coal-fired stations and felt that the Magnox reactors were an unnecessary and potentially futile diversion.

After Suez and the massive expansion of the Magnox programme called for by the Government, Brown and the others maintained their opposition, but realised that nuclear power was inevitable and sought instead to reduce the amount that the CEGB was expected to take (Hannah, 1982, pp.228-230). In this they received help from an unexpected quarter: Hinton had been appointed to the Chairmanship as a symbol of the commitment to the nuclear programme; he had indeed played perhaps the major role in establishing the physical reality of nuclear energy with his Northern factories. Upon taking office it became clear to Hinton that the large programme envisaged was out of proportion to what was required and was more than the CEGB, or the country, could reasonably afford. In 1958 Hinton told Mills, the Minister for Power, in plain language that the programme was too large, not commercially viable and contained too many risks for an untried technology (Hannah, 1980, pp.232-246). He reiterated the need for some nuclear power, he and the government whittled away at each other. It was Hinton who persuaded Mills to appoint a Board of functional specialists rather than generalists (Hannah, 1982, pp.241-246) and he used the technical findings and advice of Brown and his colleagues to persuade the Government of the wisdom of his proposal for a reduction in the planned programme, something he achieved in the White Paper of June 1960.

Brown, meanwhile, was undergoing a process of conversion to nuclear power. The early Magnox reactors returned impressive figures in terms of availability and efficiency; indeed they were run as base load, saving millions of tons of coal and oil. This convinced Brown and other senior managers that the nuclear option was one to be followed, the Board's annual reports of this era show the gradually burgeoning commitment, a commitment that reached a peak of zeal in the aftermath of the Dungeness 'B' appraisal when boasts about the low nuclear costs of Wylfa's Magnox reactor were transformed into some (occasionally outrageous) predictions about the potential of the AGR (various annual reports, but see especially 1965-66, pp.2-25).

In the years up to 1972-3, the CEGB's main reason for supporting the nuclear expansion was that the nuclear power stations reduced the ;need for what were seen as scarce and increasingly expensive fossil fuels. It repeatedly complained about the NCB's inefficient pits and called for itself to be freed from what it saw as an unreasonable support for coal, arguing that it was an independent public corporation and ought to be able to pursue its mandate of the cheapest electricity as an inherently rational and desirable course to pursue with regard to fulfilling its requirements. Brown therefore saw his mission as first consolidating the 'revolution' in larger sets and technical innovation that expanded efficiencies, and secondly to seek a modest expansion of nuclear power. Hawkins inherited the nuclear programmes of Brown and immediately sought to have them reduced, a stance he was to hold for less than a year.

Upon ascending to the Chairmanship, Hawkins claimed that some research he had been conducting immediately prior to his promotion purported to suggest an over-commitment to nuclear power. In evidence to the Commons' Science Committee he argued that in 1972 the downward revision in projected growth demand meant that the Board did not envisage requiring more than one new nuclear plant in the near future, if even that (Williams 1980, p.192, also H.C. 444-ii). Hawkins's mind was quickly changed by the miners' strike of 1972. The Board vehemently denounced the picketing of power stations and the resultant interruptions of supply; a real anti-NUM stance crept into the rhetoric of the Board and more importantly into their policy calculations too (Annual Report, 1971-2, pp.1-36). The Board perceived coal to be an insecure fuel supply and bracketed it with oil as being strategically vulnerable. From this period, particularly after 1974 and further miners' disruption, Hawkins and the CEGB sought a large expansion in nuclear power in order to safeguard the supplies of electricity (Annual Report, 1975-76, p.50). At times it appeared that the CEGB's reason for seeking nuclear expansion after 1974 rested solely on the claimed need to thwart a future strike in the coalfields; indeed one executive argued that with the projected fall-off in North Sea gas production the Board could see that coal would again be required for the production of gas and this was a prospect from which they recoiled, considering that to put both gas and electricity production under the control of the NUM would be sheer folly; nuclear expansion appeared the only alternative.

It is clear that the engineering expertise of the CEGB is interpreted in political terms at Board level and that political interpretations are transmitted down through the hierarchy for the engineers to provide technical solutions to the Board's problems; most of those solutions since the mid-1960s have been to call for more nuclear power. The character of the Chairman does, it appears, influence the workings of the CEGB, Hinton establishing the centralised machinery begun by Citrine (despite lip-service to the need for decentralisation); Brown pushing through with the expansion and modernisation of plant, whilst Hawkins consolidated the massive expansions of the past. Hawkins realised the need for large staff reductions in the light of the requirements of the new stock and these he achieved effectively and without industrial relations problems (Annual Reports, 1971-72 to 1977-78). This allowed England, a planner by training and temperament, to seek to establish the needs of the CEGB into the twenty-first century (a move that clashed with the short-term horizons of politicians), and it was he who desired that Hawkins's plans for a major switch to LWRs should become a reality, albeit not in quite the timescale or number of megawatts. A commitment adhered to by Marshall, who also re-organised the CEGB in order to insulate the nuclear component from the rest of the utility more than before (1985-6, Annual Report; Kemp,1986,pp.335-347).

The Chairman represents the views of the Board but he remains primus inter pares; his vote is only worth that of any other member. It is noticeable that different Chairmen have attempted different styles of leadership in order to achieve their goals, each seeking to break the policy inertia of a large organisation in order to pursue their ends. Each instigated a reassessment and tried to set a slightly different course and to an extent each was successful. As was seen above, Hawkins changed course after his first year and the way he attempted to impose this was through a reliance upon the centralised machinery of the Board, although careful always to base his decisions upon the advice of his technocrats. His successor, England, felt that this had led to a rather rigid structure, although on paper the structure of the CEGB has not changed a great deal over the years; it is the way in which the Chairman interprets the structure that affects the workings in practice. Under Hawkins, recalled a former Board member:

> "In practice, those who were supposed to make and carry out decisions refrained from doing so unless they knew that the Chairman was with them..... there was a large amount of centralisation and those who on paper were responsible for taking decisions frequently did not... (England) insisted upon more decentralisation; when people came to me to make a decision that was within their own competence I told them to go away and make it themselves".

This is yet another example of the importance of personal assessments within the British decision-making process as seen above in regard to the AEA and its relations with Whitehall and returned to below when the CEGB/Whitehall situation is examined. The personality of the Chairman is the important factor in his dealings with the Board, yet the political situation and the political backing the Chairman receives are also essential; for example, England lost the political support of ministers, whilst his successor Marshall, is bathed in it. At Board level the Chairman is the major interface with the political system and it is through him that the major political decisions flow back to the Board. Although his interpretation is vital, it is joint interrelationships that, according to one executive, 'work out how to steer through whatever path is decided upon'. This has traditionally been aided by the fact that the CEGB 'has always been an engineering body with a specific end product" and has therefore 'always known where it is going', in the view of one veteren Board member.

The CEGB and Whitehall

Like the AEA, the CEGB's Board is appointed by the Minister responsible for the sponsoring Department: until its inclusion in Mintech in 1969 this was always the Minister responsible for fuel and/or power. Not until comparatively recently has the CEGB enjoyed the same status at Whitehall as that of the AEA; indeed, in the early years the views of the Authority prevailed over those of the Board and the reduction in the nuclear programme of 1960 represented a compromise that was weighted in favour of the AEA. As in the case of the AEA the relationship with Whitehall has gradually evolved; unlike the entropy of the Authority's influence within the political arena, that of the CEGB has increased to the extent that it can now be considered the Government's senior nuclear adviser, a point underlined by a former chief scientist and Chairman of the AEA, Lord Marshal, being appointed to head the Board in 1983. There has been a transition in the role of the AEA and the CEGB and it is clear that the AEA's role as the Government's chief nuclear adviser is over.

The lessons of the AEA are repeated in the CEGB's relationship with Whitehall in that it is the informal networks, rather than the formal relationships, that are the important consideration in the decision-making process. Formally the CEGB is a public corporation established by the 1957 Act of Parliament. Its Board is appointed by and responsible to the Secretary of State at the head of the relevant Department which since 1974 has been Energy. Through him the Board is accountable to Parliament and thus the electorate; the accounts are audited by the Auditor General and must be reported to Parliament annually. Unlike the AEA it is not dependent upon a Parliamentary vote for its funding (being entirely self-financing), although it must obtain Ministerial approval for large expenditures and borrowings. Under the Morrisonion principles that the nationalised industries are expected to operate (see Greenwood and Wilson, 1984) the Minister is responsible for the overall strategy of the Board, but not the day-to-day operational matters; these are the concern of the Board. Constitutionally, although the Board is expected to consult the Minister on matters of importance and is ultimately subject to his decision, these are rather vague concepts and there are no detailed rules laid down for the functioning of the relationship; rather is it a matter of individual style and evolving precedent, each Minister and Board preferring to establish their own arrangements according to the prevailing political situation. In practice this rather sketchy framework is supplemented by a complex and sophisticated set of informal arrangements in which policy initiatives emerge to be the subject of intense formulation procedures.

Former CEGB executives, in arguing that because of its operating experience the CEGB was more qualified to give advice to the Government on nuclear matters, explained how the informal methods operated, the formal ones of course being minuted meetings and official documents. The informal contracts occurred when every month for example the Chairman would meet informally for lunch with the Department of Energy's Permanent Secretary. It was essential that they understood each other. In between they would also meet if things cropped up, and the Chairman's deputy would meet the Deputy Secretaries, as indeed he would too. He also had a lot of access to Cabinet Minister 'although not always as much as he would have liked'. There is no available evidence to suggest that such practices are no longer current.

Board members also maintained a network of contacts with backbench and opposition MPs in order to seek to influence grass-roots political opinion. Whereas Hinton also had direct access to Cabinet Ministers because of his tremendous prestige, in the early years the organisation he was representing lacked the influence of his old employers in the AEA, something he found irksome but which illustrates the contrast that twenty years of practice made to the respective roles of the AEA and CEGB (Hannah 1982, pp.194-296) and the way in which they were perceived by the Whitehall machinery as a whole. That is, although Hinton could dominate a single Minister the broader structure remained wedded to the advice of the Authority. The style of informal networks outlined above is a common feature of the Whitehall scene it is the quality of advice proffered and the personal appraisals that individual civil servants make of the personalities involved that decide the atmosphere in which a particular organisation operates and whether its advice is acted upon or not. In the case of the CEGB the constant striving for a nuclear component after 1974 was couched in terms readily accessible to Government sensibilities; that is, the CEGB argued it would create jobs, boost exports and frustrate a miners' strike, as well as provide cheap electricity that would aid the competitive costings of British industry. Such a scenario was bound to elicit a favourable hearing in the Whitehall circles where advice for Ministers and Cabinet is distilled and framed.

Despite the close informal networks with the Government, the CEGB has consistently sought an 'arm's length' approach from the Government, both in terms of reactor decisions (Williams, 1980, pp.244-245) and the general policy-making procedures. There has been a general cynicism on the part of the technocrats towards the politicians stemming from their belief that the decisions affecting generation are technical decisions and should be taken on those and not political grounds:

> "I always find each successive Government worse than its predecessor! Whatever colour it happens to be. It is just one of the unfortunate things of this life, that Governments get more interventionist as time goes on. It's very difficult to stop them from doing it "

argued one senior member of the CEGB who felt that the role of the Board was to make the decisions within the framework of broad political strategy and not merely to ratify the decisions of the Cabinet regarding individual reactor decisions.

The latter activity has always been a highly political decision and as the next chapter shows, there has been no decision that can be said to have been truly free of political pressure; most have been taken by the Minister. One Board member argued that although the 'arm's length' approach is the preferred one there is of necessity a:

> "...day-to-day relationship with Ministers and civil servants. One tries, as one must, to try and influence opinion formers. You talk to MPs, as we do, you try to give them the facts to

influence their thinking (you have to be careful how you do that, obviously, otherwise you can be in trouble that way). The media are not very helpful; it's bad news that makes news".

This Board member, in attempting to explain the way in which the CEGB influences policy, also noted the feelings of inferiority vis a vis other professionals (see Chapter 3), that engineers experience and attempted to use this as an explanation for some of the occasions that politicians have failed to be impressed by the technical argument:

> "We try to put our case across but there is this distrust of the professional experts; people are distrustful of experts, of scientists and engineers, so maybe they are not always the best advocates of what one is trying to do. This is what I think gave rise to the old axiom that experts should be 'on tap but not on top'.

It is because of these considerations and perceived political implications that the CEGB attempts to anticipate political reaction and to aid it in this it maintains its network of informal links, connections it also uses to influence opinion and inform a wide spectrum of policy arbiters.

It is clear that the DEn and others are dependent upon the experts for advice and the experts are not slow in recognising this dependence and framing the information in such a way that it suggests the decision that they seek. A former Chairman of the SSEB noted the technical ignorance of the Civil Service and of Ministers, arguing:

> "You have to bear in mind that the Civil Service is not very scientific or technical or numerate. Indeed, for many years out of forty or so Permanent Secretaries there has not been a scientist or engineer, so they tended to make decisions based upon broader considerations, like who was likely to win, what was the effect upon the PSBR, things of that sort. The Civil Service were reactive and certainly not originating".

Even allowing for apparent cynicism, it can be seen that the recommendations of the Fulton Report were not implemented as far as nuclear expertise and its understanding within Whitehall are concerned. Civil Servants remained ignorant of all but the broadest considerations, the gifted amateur retaining a grip upon the decision-making machinery. The SSEB man recalled:

> "I met one Permanent Secretary, had lunch with him. He said, 'Tell me, why don't we just change over to PWRs?"
> I said, 'Do you know what PWR means? And what difference is there from an AGR?'
> 'Yes', he said, 'PWR is American and AGR is British'.
> I said, 'Very good, but what about the safety case?' I asked him about the UK content and the reliability of the thing.

'Oh good heavens! I don't know details like that' he said. There's very much a political viewpoint, a luxurious viewpoint, which I think the civil service indulge in. They're committed to this notion of the gifted amateur"

As argued above, this is the technocratic view of democracy, an elitist view in many senses and one that Merton would recognise; it represents a desire for decisions to be taken on technical grounds where doubtless the technocrats feel they would have even more influence upon the decision taken. As it is they provide the technical data and recommendations and then are aggrieved if their advice is ignored by those who lack their technical knowledge on 'political grounds'. This of course is at least one view the CEGB and AEA share in common. This is to fundamentally misunderstand (or perhaps regret) the essence of the political process in a liberal democracy such as Britain where essentially issues are political if and whenever the politicians say that they are. It is the legitimate function of ministers, civil servants (acting under the instruction of ministers) and MPs to intervene, meddle and comment upon whatever they feel moved to take issue over; indeed the redress of grievance and ministerial accountability are based upon this fact. It is a feature of professions and professionalising occupations (see previous chapters) to seek to deny and evade this politicisation of issues they feel fall within the realm of their expertise, as chapters 2 and 3 show, writers as diverse as Friedrich, Habermas and Wilding have attempted to explain how this leads to a reduction in the democratic process of decision-making when it has occurred. This is not a process confined to the nuclear arena, but can be found in local government (Cockburn, 1977; Dunleavy, 1980b) and the quasi-governmental sectors (Wilding, 1982; Saunders, 1983).

Within the Government circles there is a recognition of the dependence upon the technocrats for information and advice, but also a determination not to base decisions solely upon technical recommendations. A senior DEn civil servant argued that his department's scientific and technological input comes from a wide range of external sources and varies according to the policies. There is access to CEGB, BNFL, AEA, BNOC, NCB, NRPB and a mix of scientists, engineers, accountants and other experts. It is the Department of Energy's job to look at any proposal in the light of a policy content, that can be economic, industrial, regional, European, atomic or commercial in its implications. All policies fit into and have an effect upon everything else, therefore it is the DEn's job to assess this.

This can involve Byzantine considerations and connections. It can be argued therefore that policy is shaped and moulded in this plethora of contacts and emerges in one of several ways, as a set of positive proposals, as a political directive or as a reaction or response to problems. Examples in that order being the decision to reprocess foreign fuel, build the SGHWR as part of Britain's third reactor programme or to hold a public inquiry. Thus engineers and scientists fit into a complex policy matrix, frequently supplying options for the politicians to arbitrate upon, often using their superior technical knowledge to influence the decisions, but also sometimes being disappointed at the result. As a Deputy Secretary argued, the technocrats provide the information and the raw proposals, but it is the administrators who relate those policy proposals to the wider economic and political issues, the administrators who funnel and distil the information to ministers and they use different criteria in each case.

The way in which the technocrats can have a profound influence upon the process is to court the favour of the ministers and civil servants. If they can show themselves to be trustworthy, politically astute and in sympathy with the policies of the Government then their options are more likely to be the ones adopted. A former under-secretary at the DEn recalls policy as being the process of working hard at extricating oneself from the constant muddle of lots of conflicting pressures and decisions. 'We never actually sat down and consciously made policy; rather the problems were thrust upon us in large numbers and we were then lobbied or sought advice accordingly'.

Because of the pressure of work, 'trusted' advisers would tend to have their advice taken at more or less face value, particularly if they presented it in terms of the political realities faced by ministers. A senior civil servant argued that technical people here had impact upon policy because the experts in the department would scrutinise this advice and act upon it and this gave the Chief Scientist and his department a lot of influence. It was noticeable how important individuals were; when the individuals changed so did the nature of the advice and very often the decision-making too.

Two factors are apparent here: first the importance of personal relationships within the Whitehall village, already commented upon; and second, the importance of professional relationships. Fellow professionals tend to accept each others' advice (see the last two chapters and Kuhn, 1970) so that if an engineer or scientist tells another that something should be done in a certain way to achieve a desired result then they tend to be believed, particularly if they are known and their judgement trusted. An example is the relationship between one former Chief Scientist and Marshall; the Chief Scientist recalled that when he was in Whitehall:

> "I had the most tremendous help from AEA people whose advice I could ask and who were under firm instructions to answer me as they saw fit and without regard to AEA policy. Also Walter Marshall and I have been friends for many years and he's a very honest person, a tremendous character".

This close professional relationship existed between many of those advising ministers on technical matters, thus, despite the frequent calls for a 'Morrisonion approach' the contacts are close, complex, personal and vital for the policy-making process. It can therefore be argued that politically astute technocrats who use the wealth of informal networks are able to present their policies in terms attractive to the policy arbiters. Occasionally these contacts break down as when a new government takes office. One senior manager from the CEGB recalled his first contacts with a new minister with some acute pain as he remembered how out of step they had been concerning the running of a new project:

> "I was called in to see (the Minister) and he told me I had to do this and I couldn't have that, and there were all his civil servants around him pulling his strings. I said, "Fine, in that case you build the power station and I'll send a team in to look round and if it's all right we'll buy it". They hated me for that. By the time I got back to HQ the phones were hot. The civil service told me never to speak to a minister like that again or I would never get appointed for a second term; that's how they work."

Apart from the obvious dislike of the method of the Minister, this illustrates the way in which the political process can apply very direct and severe pressure outside of the formal framework. Technocrats are kept aware of their position of political inferiority and this impresses upon them the need to maintain their networks and their position of grace. Should this be lost it is of paramount importance to repair the damage and an individual in political purdah is a liability.

This fuels the sense of frustration and cynicism felt by some technocrats towards the political side of their position. One of them was blunt about the pressures:

> "Statutorily you are responsible (for the actions of the Board) but all the time you are under pressure from civil servants, from government. There is always the threat, if you're not reappointed you've got nowhere to go; you can't just get back into the system. Which is what they did to Glyn England...they didn't give him more time and I think that the way that they handled it was disgusting".

Informal pressures of this type are a frequent and effective method of keeping the technocrats in check; Hawkins was another senior figure who felt this discipline in some considerable measure.

The sophisticated nature of the relationship between the CEGB and Whitehall is apparent. The formal structures and relationships are but a veneer behind which the real business is conducted. The technocrats possess considerable power in the form of their technical expertise and specialist knowledge; this allows them to frame the policy options which are then subject to the constraints of the prevailing political situation. Politically astute technocrats are able to use their political experience in tandem with their technical expertise in order to frame options in such a way that they will appeal to the civil service and politicians, political experience and knowledge being acquired from a constant stream of contacts. Examples of the subjective framing of policy options form the basis of the following chapters, although these deal as much with the AEA as with the CEGB. The CEGB has been able to supplant the AEA as the Government's nuclear adviser (de facto if not de jure) because its experts have been more successful at cultivating the correct

Whitehall relationships since the 1970s, a result of the immense experience and mature political awareness of the Board. As is shown in Chapters 6 and 7, the CEGB since 1979 has expounded goals that have been of some considerable appeal to the Conservative Government; this has obviously added to its influence and carries on a tradition of government favour (the technologically-minded and managerial ethos of governments of both parties in the 1960s and 1970s were also attracted to the CEGB's style) that has supplanted the AEA. The belief within some influential circles, that the advice of the AEA has been less than adequate has hastened the fall from grace for that organisation.

The CEGB and Private Industry

The relationship between the CEGB and private industry has at times shown a marked contrast to the cordiality that has prevailed between the AEA and the consortia and contractors. Whilst peace, if not tranquillity, has reigned in the dealings between the Board and some firms, with others the Board at times operated in a state resembling industrial warfare. This was the situation in the manufacture of both nuclear and conventional plant, the situation at the Isle of Grain oil-fired plant and Dungeness B AGR, being just two examples. Much of the analysis regarding this facet of the technocratic role in policy-making is located in Chapters 6 and 7 below, as it is concerned with the reactor decisions; the relevance for this chapter is that the CEGB has considered itself (and is so considered by its employees) as the premier employing organisation within the electrical and associated professions. The links between the engineers within the CEGB and those employed by private industry are such that the CEGB engineers have exercised a hegemony over their private sector counterparts - a situation that tends to confirm the hypothesis of Dunleavy (1979) who argued that state-employed professionals may come to focus or supervise the work of private practice or corporate sector professionals on a quasi-managerial basis and that where the work, research and development function, and non-routine application of knowledge are concentrated, as in the CEGB, then the profession's cultural system will tend to be situated (1979, pp.13-15). Unlike in professions such as architecture, the private sector has not taken on the role of an elite, the United Kingdom lacking the large independent firms that exist in the USA and work as architect-engineers designing and supervising the building of large engineering projects. It must be noted, however, that the management of large firms, such as GEC, do come from a predominantly engineering background.

Williams (1980) and others have documented in some detail (cf. Ince, 1982; Pearson, 1981; Hannah, 1982) the evolution of the consortia system for building the nuclear power stations. The CEGB at first welcomed the system; Brown in particular insisted on retaining the consortia, arguing that it injected the discipline of market competition into the tendering arrange-ments. The CEGB made it clear throughout the 1960s that any attempt to reform the consortia and consolidate them into just one organisation would be met by the CEGB ordering plant from abroad in order to ensure competitive pricing. Thus, although both the consortia and the AEA wanted to rationalise the industry, realising that the United Kingdom lacked the domestic market necessary to support the existing structure they also realised that they could not afford to risk the CEGB inviting foreign tenders. The threat of foreign competition was one that from the early 1950s the ESI had periodically wielded in order to discipline the private sector, but as Hannah argues, it is questionable whether the Government would have allowed the bluff to become a reality (1982, pp.116-117); indeed the engineers within the ESI themselves were reluctant to use this weapon against their colleagues in the private sector, even when political clearance was given in principle in the mid-1950s. The Sizewell 'B' PWR is no exception, although an American system, the design, construction, and most components are and will be tendered to UK firms.

Despite the insistence upon retaining the consortia, the CEGB did not in fact always accept the lowest bid, but operated a Buggins's turn in both nuclear and conventional plant, a system which infuriated the Select Committees in 1967 and 1980. Often the CEGB (and its predecessors) would intervene to bail out the weakest firms and use their own engineers to support their technical people, all in order to retain the fiction of competition. Members of the AEA described Brown as having the mentality of a city engineer, a charge echoed by Members of Parliament, both politicians and the AEA arguing from the mid-1960s that there could be no true competition in the nuclear sector and what was required was the establishment of a strong industry capable of meeting domestic demand and exporting abroad. This could not be achieved with too many firms chasing too few orders. Eventually the CEGB also came to realise this, but it wished the industry to be established on its terms; it was fearful that companies with which it had experienced a difficult working relationship, particularly GEC, would dominate the industry. As a SSEB executive pointed out:

> "The CEGB has always been big enough to call the shots. They've been able to dominate in the way that British Airways were."

The CEGB's research and development budget dwarfed private industry's and their expertise at Barnwood and elsewhere was a match for anyone; the Board was able to insist upon its way and to check up on the work of private industry.

Much of the bitterness between the CEGB and private industry came from the nature of the turnkey contracts used for the nuclear stations and some of the larger conventional ones, such as Ince, Isle of Grain and Ferrybridge. Because of the competitive system (much modified by Buggins's turn) the consortia would quote low prices to obtain the contract and then the CEGB would have to buy them out and finish the project itself using its own people and sub-contracting out for specific items. Dungeness B is the prime example, being nearly two decades late in construction; but there were also long delays in Hartlepool and Heysham. A CEGB engineer who was responsible for sorting out the morass at Dungeness and Isle of Grain, recalled that after APC had been bought out at Dungeness he sent his men in:

"But when we got in amongst it we found all sorts of things had been done that didn't stand up to the types of tests and the criteria that we wanted. So quite a lot of it had to be ripped out and back-filled."

It was not just APC that had caused substantial problems for the CEGB:

"I think it would be true to say that neither of the generating boards were enamoured with the BNDC/Whetstone outfit. What concerned many of us was that if the industry was rationalised it would come under Weinstock and we didn't really like that much."

The Board wanted the industry to operate the way that the other major consortium (TNPG) had, through a series of hard contractual agreements linked to a co-operative approach whereby CEGB specifications were incorporated into the work. A Board member recalled:

"If there was a problem it was discussed with our scientists and engineers. I think you've got to bear in mind that the CEGB spent something like £80M a year on R. & D., covering everything from fatigue through to nuclear power and acid rain. Whereas the NNC was spending something like £.5M. For example, the CEGB would spend more on non-destructive testing than the NNC would spend on total research; the net result was that the CEGB always had first class experts.

Because of their jaundiced view of private firms, the CEGB effectively took direct control of the manufacture of their power stations, arguing convincingly that turnkey projects had turned into cost-plus projects that had cost the Board dearly, yet short of bankrupting the private consortia, there was no redress. One executive pointed out that:

"If the Board had stayed with those contracts and had enforced all the penalties, at the end of the day they could have broken GEC, NEI, Babcocks, the lot. But that's not a sensible thing to do."'

Many former CEGB managers reflected bitterly upon the role of some of the private firms, considering it a failure of capitalism in that the private sector simply lacked the resources to do the job and had to be directed. An ex-Chairman argued:

"You must always remember that the function of private industry is to make money; I sometimes frustrated them because that is not the function of the CEGB."

Another former Board member bluntly pointed out that although it may be 'a little harsh' to say that private industry wanted the AEA and CEGB to bear all the research costs whilst it took all the profit, it does 'contain a lot of truth'. He said that unions were bribed into good behaviour with the costs being passed on to the Board and thence to the consumer:

> "The truth is that GEC and others promised to supply us with power stations by a certain time and at a certain price: they did not do so."

They were however able to avoid the penalties normally associated with such a failure and although the private manu-facturers may be smug about this escape, the CEGB's technocrats developed a lasting enmity that became a part of their world view. Some argued such an attitude is the difference between private industry and the CEGB, the Board seeking to serve in an ethical sense very closely resembling the ethos of the social professions (see above Chapter 2), wishing to build power stations that will benefit the community and will last for twenty or thirty years, whilst 'the manufacturers just want to collect their money'. The Board was anxious to ensure that industry be made accountable, if not through competition then through enforceable penalty clauses.

It was perhaps inevitable that the Board, in developing the public service ethic typical of professions, would come sometimes to sound sanctimonious. Through claiming that only it had the public interest at heart and therefore had to be the final arbiter of what it wants, clashes with private manufacturers became unavoidable. The most obvious reason for this was that Britain was the showroom for British products; it was in the domestic market that manufacturers hoped to develop products that they could then export. The ESI understood this well, but whereas the SSEB was willing to compromise, the CEGB maintained that it would seek only those products it felt were required for the British market, irrespective of world demand. This was criticised by some in the SSEB who argued that it was damaging to British industry for the CEGB to behave in this way, one executive argued:

> "When I was Chairman of SSEB we did at least change designs in the way the GEC acknowledged that they could sell abroad." (op.cit.)

The CEGB, however, denied the charge and pointed out that the utilities across the world are 'a little club' and 'we all know each other'. Because of this even if the CEGB buys a particular piece of equipment they are likely to discuss its applicability elsewhere with other utilities. One senior executive cited the example of when the Board decided not to purchase a particular set of turbines from GEC; the South Africans were about to cancel their order, too (and thereby fuel industry criticism), but the CEGB pointed out that the order they were about to cancel was the optimum range of equipment for them. As a result of this consultation they placed their order with GEC. It is not possible to judge the extent to which this is an isolated example, but certainly industry felt that the Board could have done much more to aid them, whilst the Board remained firm in the belief that it was industry's job to find its own orders. In the event that is what firms like GEC have done, GEC itself now ensuring that most of its business is centred in other EEC countries and the developing nations. It is interesting to note that in the lobbying for the PWR, the CEGB has used some of the arguments of the private sector, namely that if the CEGB adopts the PWR it will be easier for the private sector to export components.

The Board, in sticking to its mandate of providing adequate supplies of the most inexpensive electricity, antagonised much of the private sector. It in turn found the standards of the private sector to be inadequate and to the charge that it did not have to make the same profit, pointed out that its mandate was as severe as the discipline of the market and that the nationalised industries were accountable to the commercial leanings of ministers who have increasingly insisted on a profitable return for public investment. Certainly the CEGB has seen in the criticisms of private industry attempts to make a profit on the back of public investment and viewed GEC attempts to close down Barnwood as a ploy to cripple the ESI's ability to effectively oversee the often shoddy work of the private sector in power station construction. The Board accepts the need for private industry to make a profit but has insisted on a greater degree of commercial accountability and discipline than has been available in the restricted market of the UK; in many respects the strictness of the CEGB's requirements have acted almost as a surrogate market, providing a measure against which the private sector has often been found lacking. As one ex-manager explained to one of Mrs. Thatcher's Cabinet:

> "I am in favour of private industry, I have a lot of contact with it. What I don't like is the kind of private industry that we have at NNC and which you lot all rave about just because it's private. I think that some things must be run publicly because it's too dangerous to try and extract profit from them; nuclear reactors are an example. I would not let ..... anywhere near them; he's the sort of private industry that I don't like."

The CEGB's engineers therefore consider the aim of private industry to make a profit perfectly respectable and account for it, but in their 'professional' opinion there is a higher concern; the public interest. Thus although they have always taken a pride in the development of the commercial ethos within the Board, particularly from the early 1960s, seeing the demand for cheap electricity as a strict 'market' form of discipline, they have also felt it essential to keep a close check upon the more profit-orientated private sector, which has not shared the same commitment to the general welfare. This has accounted for the antagonism over the management structure envisaged for the Sizewell B power station, which some saw as a government enforced return to the old and discredited methods of the past. But the CEGB took the opportunity of the Public Inquiry to review its position, particularly after the evidence of Sir Alistair Frame, and it became clear that the Government was prepared to bow to the lessons of the past and allow the CEGB to manage the project and put the components out to tender instead of relying upon a joint management board with NNC (Kemp, 1985; Lomer, 1985; Layfield, 1987).

The CEGB and the AEA

It was claimed in the concluision to Chapter 4, that the CEGB can be viewed as a Rome to the AEA's Greece. Much of the evidence for this is contained in the next chapter, or has been discussed above; but the main points are that the two possessed different worldviews; the CEGB in absorbing the expertise of the Authority came to possess a greater and more commercially relevant expertise; and relations between the two were somewhat sour for much of the 1960s and 1970s.

It would be tempting to ascribe the differences in worldviews between the CEGB and the AEA to the contrasts in the ethos of scientists and engineers. This would not however take cognisance of the fact that there existed within the AEA itself this same contrast and although there were tensions present, it never acquired the bitterness that was to dog AEA/CEGB relations for many years, even after the introduction of over-lapping board membership. The next chapter illustrates how much of the tension lay in the process whereby the CEGB felt it was subject to the atomic preferences of the Authority which, because of its vested interest in 'its' reactors tried to foist upon the Board whatever the AEA felt the ESI required. This indifference to the CEGB's wishes was to become a recurring complaint of Hinton's and later Chairmen realised that although obviously some misunderstandings occurred due to the engineering bias of the CEGB and the research function of the Authority, much of the difficulty lay in the pursuit of different goals by the two organisations which was as a result of their different functions. As one explained:

> "AEA goals were scientific and research based, and CEGB ones were engineering and commercial, but it was organisational not professional matters that led to the clashes."

The relationship between the two was bound to deteriorate as the AEA fulfilled its function of expanding nuclear knowledge and educating the ESI and the industry. Once it had fulfilled its basic research and training functions the Authority was left without a clear role, whereas the CEGB was always safe in the firm conviction of its position as the largest part of the ESI. One of its first nuclear engineers recalled how:

> "...the AEA had been set up to find out what nuclear energy was all about, to build the bomb etc. It realised it had achieved that goal and then it wanted to bring it into the civil structure through Hinton. So it started building the infrastructure for the civil programme...they had flowed onwards in a creative vein and the AEA was left as a decaying appendix from the 1960s; its function was over but it still stayed there."

In the view of the CEGB the master/pupil relationship had by the mid-1960s become one of peer equals and then they perceived the AEA as a superfluous entity. This was particularly the case after 1971 when BNFL and Aldermaston were hived off, the CEGB retaining cordial relations with the commercially-minded engineers of BNFL, although critical of some aspects of their behaviour.

The relationship has then been a dynamic one with the CEGB starting as the novice before assuming the mantle of nuclear expert. In the early years the CEGB was not expert enough to challenge the AEA; only Hinton could query the thinking of his former colleagues. These were organisational differences, Hinton realising when he transferred to the CEGB that many of the engineers' claims to have improved the efficiencies of new coal-fired plant were indeed true and reduced the claimed superiority of nuclear economics; indeed the advantage slipped back from nuclear to coal. Later differences were of a more commercial nature regarding reactor choice. Despite the organisational differences, relations between the individual scientists and engineers remained cordial, even close; each felt free to wander through the others' buildings and seek advice upon any subject; indeed it was the professional contacts that retained the links between the organisations. These links were particularly close between the CEGB and the Production Group engineers, since although the two performed different functions they were dependent upon each other.

The CEGB in Perspective

Although not operating in a market in the classical sense, the CEGB is a profoundly commercial organisation. It regards its mandate to supply electricity as and when demanded at the lowest possible price as a strict discipline fulfilling the role of a surrogate market; indeed its careful operation of a merit order of its stations is a finely-tuned operation designed to protect the consumer. Linked to this commercial awareness is an acute sense of public service reminiscent of the public service ethic of the social professions, and the dominance of the engineering groups is a plausible explanation for a service ethic in the CEGB. Because of its professional stance and because of its manifestly important role, the ESI is seen as the dominant institution within the culture of those engineers associated with the industry, the CEGB's engineers acting as managers to the private sector and exercising an elite function. The dual allegiance of engineers analysed in Chapter 3 is in evidence within the ESI.

The CEGB has supplanted the AEA as the Government's chief nuclear adviser, in fact if not in law. The complex network of informal contacts acting as a conduit through which the experts provide the policy options upon which the civil service and government arbitrate. The workings of the networks are a valuable lesson in the way in which policy originates, is formulated and distilled into action within the British political system. It is also a lesson which the technocrats have had to learn in order for them to present their options in such a way that they sway the decisions in the direction they would like.

The Technocrats and Policy-Making

It can be argued from the evidence presented above in this and the last chapter, that the policy implemented in the field of civil nuclear energy is the result of myriad separate political and technological factors of which the technocrats are just one, albeit important factor. In his evidence to the Select Committee on Energy in 1980, Benn identified almost every office in the Cabinet as having a legitimate and continuing interest in nuclear policymaking (H.C. 114-ii), a point he expanded upon in his evidence to the Sizewell 'B' Inquiry (1984). The obvious importance of the policy to the Departments of Trade, Employment, Industry, Defence, Environment and the Treasury, Home Office and Foreign Office, meant they had a close interest, whilst the Scottish and Welsh Offices, Social Services, Transport and Consumer Protection also sought to influence decision-making. As the arguments of the civil service analysed above made plain, decisions are taken on political grounds and the weight attached to technical arguments varies according to the policy under consideration; this is a theme that is clarified by the analysis of subsequent chapters.

The influence of technocrats lies in their ability to relate the political requirements of the elected politicians (and their civil servants) to the technological realities of which they possess the expertise. Thus, policy is a dialectic between perceived strategic national goals and the technical options available to obtain those goals. The frustrations that break out on both sides in this process can be glimpsed in a frank exchange between Palmer and Marshall during the presentation of evidence to the Energy Committee in 1980, when Marshall said that certain questions of reactor safety should be left to the professionals and should not be taken by Ministers. He said:

> "I see no reason why we should not rely upon that type of...technical decision-making".

Palmer replied:

> "I can see all that. The point is that this is a democratic government. Governments are elected." (ibid)

MPs clearly felt that the technocrats' attempt to circumscribe the areas of legitimate political interest to be itself an attempt to interfere in the decision-making process and abrogate authority.

Benn himself, as the minister with the greatest experience of nuclear power in Western Europe, symbolises the growth in the confidence of politicians when dealing with the technocrats in fields previously felt to be the preserve of the expert. By 1980 the confidence Maudling had expressed in the scientists of the AEA in the early 1960s had evaporated. Whereas in 1967 Benn told the Select Committee on Science and Technology that, although he would not give work to the AEA just to keep the teams together, 'but only on the basis of clearly-defined need'; he also admitted that Penny and the AEA were his principal advisers on atomic energy and it was 'not right or proper' or necessary to duplicate the AEA or to review its advice (H.C.381-ii, 11/5/67). By 1980 he was to argue that the Energy Minister found it hard to get the information he needed from his department:

> "I think there is a deep institutional failure which affects ministers marginally, Parliament in a major way, and the public substantially, that makes it very difficult to establish democratic control and parliamentary control over these matters" (H.C.114-ii, pp.414-419)

The admission was itself a major development: previously politicians had been content to admit and accept their ignorance. By 1987 his view had hardened to a belief in the anti-democratic nature of much of the Whitehall machinery and the technocrats (and civil servants) who informed and were a part of it.

The Energy Committee reflected the general loss of confidence in the AEA's ability to deliver its promises and perform its duties properly when it drily noted that although the Authority had been the Government's principal adviser it had been a poor adviser and should therefore concentrate upon scientific matters and research (H.C.114-1, p.71). As was noted above, although the Government's reply appeared to defend the AEA, in practice the political system had already demoted the status of the Authority's advice, something that dated from the fiasco over the SGHWR (see next chapter). Indeed, from about 1973/4 the Government was beginning to move closer to the position of the CEGB, a move only temporarily slowed by the 'hiccup' of Varley's decision for the SGHWR .

It can be argued that the relationship of the technocrats within the CEGB and AEA to each other, to their employing organisations and to the political system is in keeping with the analysis of previous chapters regarding the dual allegiance of the technical experts. Caught as they are within the bureaucracies of the large organisations that employ them, the quasi-professional occupations display a mixture of professional and organisational motivation. Within the CEGB and the AEA they dominate the hierarchies, and policy options therefore reflect the view of the technocrats - views however also shaped by organisational viewpoints which in the hived-off parts of the AEA (BNFL and Amersham) and the CEGB tend to be more commercially formed and aware than those of the science-based AEA, although there has also been a change here since the 1970s and the leadership of Marshall. Within organisations dependent upon commercial success, or which are motivated by commercial goals, such as the consortia and the CEGB, organisational goals tend to become professional goals. Whilst in the research parts of the AEA there has been a propensity for professional goals to become organisational ones. In both types of organisation, however, the two motivations merge to help synthesise policy. As nuclear power has matured a similar maturing of the political awareness of the technocrats has been evident, with options being presented to higher management and to the policy-arbiters of the political structures, in a form which recognises overall political strategy and is designed to accord with this. CEGB proposals which seek to maximise their chances of being adopted are the result therefore of exhaustive informal contacts and are prepared in the light of what the politicians are perceived to desire.

As a result, there is a decisive amount of conscious and sub-conscious screening of advice and although this advice is frequently followed (see Chapter 6) and is therefore influential, it has itself been greatly influenced by the prevailing political environment. Advice that was not so framed to take full cognisance of the minister's mind would not be successful in influencing policy. An example of the latter would have been a proposal to construct another coal-fired power station in the Yorkshire coalfield in 1979; such a proposal would not have been considered - indeed, it took a great deal of Howell's political skill to persuade the Cabinet to build the two AGRs still on order instead of opting immediately for the PWR. Such was the mood of the Government in the early 1980s that a British option was regarded as a joke in poor taste. The next chapter presents an analysis of why this change from 'nuclear nationalism' to 'nuclear internationalism' occurred.

6 The reactor decisions I

Chapters 6 and 7 should be read together as they seek to clarify the implications of the last two chapters regarding the dependence of technical professionals upon their organisational bureaucracy and the subsequent effects of this upon their professional goals. Both organisational and professional motivations are exposed (facilitating clarification) by the decision-making process of the reactor choices; this chapter being concerned with the AGR and Chapter 7 with the SGHWR and PWR. The following section below explores the reasons why the AGR was developed, before subsequent sections proceed to describe the 1965 decision in detail and examine the various organisational interests involved; concluding with a discussion of the lessons the AGR saga taught the professional experts and others involved in (or observing) the decision. This chapter represents a shift in emphasis from the earlier concern with structures to a detailed analysis of the motivations for activity, mooted in earlier sections of the thesis. It is a task that is embarked upon after the next section has performed the prerequisite of explaining precisely why the AGR came to be developed at all and how the choice was made. Whereas the last two chapters plotted the role of technocrats in the structures of the constituent organisations of the policy system, and the differing influence of the AEA and CEGB over the politicians, Chapters 6 and 7 take earlier points raised there and work through the reasoning behind the decisions from the perspective of the technocrats in order to explore that perspective and measure their success in achieving their goals.

The lessons of the reactor choices reinforce the conclusions reached regarding the relationship of the technocrats to the political system, in that the technical ignorance and dependence of the politicians is an obvious factor in the influence of the experts; but so is the importance of political considerations such as strategic energy independence, cheap electricity for industry and the potential of British exports. The chapters illustrate that a shift in commitment and perception by the policy system was as a result of the evolving technical possibilities available to the policy-makers, but also because of the need for the technology to 'pay' for itself in hard commercial terms as well as the obvious political benefits. The arguments of radical writers such as Jessop (Littlejohn, ed.1978) suggest the necessity of the Liberal Democratic state to provide the conditions necessary for the reproduction of capital accumulation on which the structure of the society is founded, and this in turn means that options with a propensity to maximise returns on the international market and minimise costs on the domestic market are those that are favoured by Government. In short, the British reactors did not sell and were expensive to build; therefore, the American alternative which had been adopted throughout the West was to be increasingly preferred, leading to the loss of the AEA's influence and the corresponding increase in that of those who argued for a more commercial approach.

Whilst a strictly chronological approach is avoided it should be remembered, as a guide, that the decisions were linear to the extent that with hindsight they appear almost as a natural progression. The technical, resource and political expediencies had led in the 1950s to the choice of the gas route for Britain and this culminated in 1965 with the choice of the AGR in preference to American light water reactors; a choice also faced by Secretary of State Peter Walker in 1987 when he opted in favour of the American system.

THE AGR

The Advanced Gas-cooled Reactor was first mooted in the mid-1950s at Harwell and Risley even before its predecessor, the Magnox, went into full commercial production. A decision to proceed on a narrow front in research was forced upon the Authority by resource constraints, it being necessary to select a small number of reactor types for development and to concentrate upon just one in a concerted manner. In the early 1950s, the Authority considered several hundred different types of reactor (Gowing, 1974, pp.36-90; 203-260; 262-302) but narrowed this down to natural uranium, gas-cooled, graphite-moderated reactors and light-water reactors. Work on the LWRs continued for several years and at one stage it was envisaged that Britain would progress from the Magnox type to the LWRs (see previous two chapters), but the lack of American cooperation over resources, particularly enriched uranium, meant that Britain had to pursue a course that utilised natural uranium.

Because any type of water reactor meant some form of enrichment, either of the water (as in CANDU) or of the uranium (and Britain's resources in the early years could not stretch to providing enrichment facilities for fuel) the LWR option was discontinued. Another factor in this was that the Magnox-type of reactor bred a lot of plutonium and this was required for the military programme. Towards the end of the 1950s enrichment facilities were available within the Production Group of the Authority at Capenhurst and this would have been the time when Britain could logically have made the switch to water reactors. It became clear however that by about 1957-8, the AEA had become committed to he development of an advanced gas-cooled reactor (the AGR) and after this selection had been made it would have been difficult to prevent the full-scale development of it. In 1960 Hinton had said that in his view the AGR was 'unquestionably the logical step forward in British reactor technology' (quoted in Williams, 1980, p.88) although just a year later he argued that it 'may be necessary' for the CEGB to purchase foreign-designed reactors and that it was not important for Britain to make every invention in the nuclear field in order to remain a world leader (Williams, 1980, pp.88-90).

As is shown below, the Government's adoption of the AGR in 1965 stemmed from the privileged position of the AEA in nuclear decision-making and its presentation of this reactor as a breakthrough in British technology that would not only bring international plaudits to add to those Britain was showering upon herself because of the Magnox innovation, but would also bring in its wake lucrative export orders. The evident success of the AEA's Magnox reactor and the satisfaction the ESI showed with the power stations it possessed that were fuelled by Magnox, plus the two export orders obtained, had covered the AEA in glory. The opinion of the Government was apparent from the way in which it permitted no criticisms of the AEA to sway it, indeed there were few such criticisms (see above chapter and Williams, 1980, pp.82-3). Reasons for the AEA's adoption of the AGR are less clear, but it can be strongly argued that the professional attachment of the research teams was the instrumental factor involved. The analysis of chapters 4 and 5 indicated that professional experts evolve a strong attachment to the team in which they are located and that they also identify with and promote the field of research upon which they are engaged; a case study of this forms the basis of Chapter 8 with regard to the Production Group of the AEA. The teams seek to have their goals adopted by the organisation as a whole, whether as in the case of the fuel cycle experts this is just one group, the Production Group, or, in the case of the AGR teams, the Authority per se. Although not all those engaged upon reactor research were committed to gas options, those under Cook at Winfrith for example preferring the heavy-water option (see below), the dominant groups at Harwell and Risley were in favour of advancing the development of the gas-cooled technology. The reasons are unclear but appear to include a variety of factors.

The period in the middle-to-late Fifties, when the decision was taken to concentrate upon the AGR, was the time when the Authority, and Britain, could have switched easily to LWRs because this was something of a hiatus. The Magnox were ordered and were being developed, but it was already clear that they were an obsolete technology (Dent, in Marshall, 1983, pp.154-155), their temperature and fuel ratings being lower than the new coal stations; whilst the Authority from this period possessed the ability to manufacture enriched fuel in its new enrichment plant at Capenhurst in Cheshire. Instead of moving over into LWRs the AEA decided to use the advantage of enrichment to advance the operating efficiencies and potentials of gas technology. It is easy with hindsight to criticise this decision; indeed, even Hill has said that Britain should have moved into LWR technology many years earlier than it did, arguing that standardisation on a worldwide scale would have brought rich rewards; yet the case for LWR had not at that time been made. Indeed, one engineer who was embarked upon a course at Harwell during this period, recalled that:

> "Nobody knew whether the answer they'd got was the right answer; everybody knew there were terrific developments going on, therefore there was a tendency to be somewhat open house, each one claiming of course that they knew the answer - but they did it with their tongue in their cheek. It rapidly narrowed itself down to AGRs, PWRs and CANDUs. Inherently that was the thermal reactor spectrum. Each one of these was dictated to a large extent by its history, i.e. the PWR was dictated in the States by the Rickover Programme...of course our own gas-cooled reactors arose out of the fact that initially the Americans were reluctant to let us have any enriched fuel. When we came to look at the PWR at that time it was evenly balanced between LWRs and gas".

The arguments propounded by the AEA's technocrats therefore appear to have been: firstly, that the LWR was even less tried and tested than the Magnox and gas technology; secondly, that since Britain had established a world lead in the practical application of nuclear technology for civil purposes, to opt at any time for American technology would be to denigrate British achievements and cast away a lead won in the teeth of American intransigence over co-operation; thirdly, that Britain had established an expertise in gas reactor technology and that too much had been invested in terms of research and other resources to switch. Finally, that as a 'proved' and established technology all that Britain had to do was to remain firm and the export orders would soon appear for AGRs.

In essence, the AEA opted for the AGR because they felt it to be a logical extension of the work they had done for the establishment of the British nuclear capacity. It also had some clear technical advantages in that it was a single phase fluid, reducing the need for so much enrichment and thus reducing costs. Burn (1978) whilst acknowledging the professional and nationalistic attachment of the AEA to the AGR, points out that when the AEA opted for the narrow front approach it made a choice of system 'with no prototype experience of alternatives' (1978, p.12) and that it should have followed the German example and developed LWRs alongside the AGR. As a result of the narrow front, he argued, the AEA missed the opportunity to realise the potential of LWRs which by the end of the 1960s were the only reactors with an export market (1978, p.96). It appears therefore that the decision to opt for AGR development owes a lot to organisational constraints of the type noted above in chapters 4 and 5; that is, the political and resource realities of the 1950s ensured that Britain began along the gas route and then, when other options presented themselves, the organisational attachment to the prevailing goals (the result of professional inertia) caused those alternatives to be foresworn.

Another reason for the early attachment to gas being institutionalised in the AGR was that no one within the AEA at the time expected thermal reactors to remain viable after 1975. Indeed, the earliest goal of the Authority was not a thermal reactor at all, but a fast breeder and that has remained its premier end up to the present; the Magnox and AGR were regarded as nothing more than stopgaps (Burn, 1978, p.2, also Williams, 1980, p.20; Penney's evidence to SCST, 1977, H.C.381-1). Gowing has shown how in the early 1950s, when the primary aim of Atomic research was the production of nuclear weapons, the responsibility for the research into power reactors was shared between the scientists at Harwell and the engineers at Risley (1974, pp.248-252). The decision to build the Magnox was taken because it was the quickest route to he rapid production of fissile material, but the fast reactor was confidently expected to proceed in three stages: 'zero-energy and lower power stages which would be Harwell's responsibility and a full-scale reactor which Risley would design'. This would lead to the full-scale reactor and 'low-power reactor merged into one large experimental reactor under Risley' (1974, p.351). Thus the thermal reactors were seen as users of the heat by-product that came from the military purpose of producing plutonium (Dent, 1983) whilst the fast reactor would become the heart of the power stations of the future.

Penney felt that 'ultimately it may be the only reactor system to be installed' (SCST HC 381-1). By the time the Government announced its decision in 1987 to build Sizewell 'B' PWR, the FBR remained a distant option with no plans for a commercial operation in the UK, although the British technocrats looked with envy to the French Superphoenix, a full-sized commercial fast reactor. The nearest the UK has come to the fast reactor dream is the Dounreay Fast reactor begun in 1955 and completed in 1959, and the Prototype Fast reactor (also at Dounreay) which first generated electricity in 1977 (Annual Report, 1977/78 of the AEA). The earlier decisions in nuclear energy should then be viewed in the light of the commitment to the FBR, a commitment that has remained strong as was indicated by the 1984 decision of the AEA to collaborate in a European project and to build a reprocessing plant at Dounreay to cope with the demands of a European programme (Guardian, 25/5/85 and Atom no.345, July 1985). The narrow front, a result of resource limitations, dictated that Britain could not simultaneously embark upon the LWR and gas technologies and the decision to opt for gas was therefore influenced by a combination of professional attachment, national pride and sound technical judgement.

The 1965 Decision

The decision of the ESI in 1965 to select the AGR as the reactor to fuel the second nuclear programme became a subject of some considerable controversy, with a strong belief expressed by later commentators that there was a degree of Government intervention which effectively swayed the decision for the AEA and its AGR (Williams, 1980, pp.209; Burn, 1978, pp.150). The evidence is confused upon this point, although there are grounds for strong suspicion that the

Government did intervene; considering the importance of the decision, it would have been strange had the informal networks, outlined above in chapters 4 and 5, not been used to inform the decision-makers of the Government's preference for the AGR. It is interesting to observe that when Varley and Benn came to make the next set of decisions on reactor choice (1974-8) they did not feel the need to utilise such devious methods, but simply considered the advice tendered and made the decision, which they then communicated to the ESI. The procedure established to make the decision is well documented in Burn (1978) and Williams (1980).

The decision on Britain's second nuclear programme took place amid a background of deteriorating relations between the CEGB and the AEA. The Board had resented the manner in which the Authority had used its position of being nuclear adviser to the Government and the ESI to press for a programme and a reactor that the CEGB had begun to have some severe doubts about. In a penetrating assessment of the various choices open to the Board, Hinton in 1961 had argued the various costs and benefits for the AGR, CANDU and LWR (Nuclear Power, in Three Banks Review, Dec. 1961, pp.3-18). It was clear that the CEGB was openly canvassing the possibility of a foreign reactor and Hinton wished to keep the industry's options open. Hinton also developed a deeper resentment of the Board's dependence upon the AEA and criticised his former colleagues for being overstaffed, underproductive and in the habit of unloading the Authority's overheads on to the Board. He argued that the AEA should have sought to become more commercial and seek a better return from the money invested in nuclear power (Hannah, 1982, pp.241-246). Questions of commercial advantage rather than simple strategic investments had come to dominate the thinking of the Board. This was a new departure for nuclear decisions, as the SC on Science and Technology had noted that the Magnox programme was founded on strategic considerations:

> "Cost was a secondary matter, either dismissed altogether or brushed aside on the grounds that, with a new method of energy production, costs were bound to be high at first but would gradually come down as experience was gained" (SCST, HC 381-i, para.99).

By 1961 the Board felt that it was time to reduce the costs, and the fact that the AGR prototype at Windscale was not even working (eventually coming onto full power in 1963) led Hinton and other members of his Board to look closely at alternatives, although they were to 'remain pessimistic about both the PWR and the BWR' in 1961-62 (Williams, 1980, p.90). In 1962 the Government sought unsuccessfully to attempt to reconcile the Authority and the Board, and appointed a Permanent Secretary, Sir Richard Powell (from the Board of Trade), to head a committee to investigate reactor choice. Unable to resolve the issue with Powell, the Government sought a way out of its impasse and published the fruits of Powell's deliberations as a short White Paper in 1964 (April, 1964, cmnd.2335). The Government decided to present the choice of reactor as a straight commercial decision decided by the CEGB acting with the advice of the AEA. It was revealed that the Board would review tenders for an AGR, and would also consider tenders for a LWR submitted by British industry.

Williams described the review as a ruse to extricate the Cabinet from the position of appearing to side publicly with one public body against another (1980, p.12), a consideration Varley chose not to emulate in 1974. Of more importance to the industry, the White Paper showed that the station chosen would be the first in the second programme that it envisaged would total 5,000MW. The apparent willingness to consider another type of reactor was not lost on the Americans. In 1963 they had announced a breakthrough in costings with their BWR at Oyster Creek (Burn, 1978, p.23-44, 96-108; Williams, 1980, pp.108-11, 134-50). The Jersey Central Power and Light Company in July 1963 invited tenders for a station, either nuclear or conventional. In December they announced that General Electric's BWR was economically superior to a coal plant of the same size and that this was 'an historic turning point' in the search for competitive nuclear power (Williams, 1980, p.110).

At that time it was accepted in Britain that the relative economies of coal and nuclear were slanted in favour of coal (Sweet, 1982,) but with the American 'breakthrough' the choice for the new reactor had to be seen to be competitive with conventional plant as well as other nuclear reactors; as Sweet shows, there was an enlightened interpretation of statistics to 'prove' this to be so (1982 and 1980). In 1965 the CEGB received seven tenders for Dungeness B; three from the consortia for the AGR, three for the PWR and one for the BWR. Each tender was broken down into 71 parts and detailed critiques written on each part by a joint AEA-CEGB team. The separate reports were then aggregated and considered by the senior Board and AEA engineers as one document (Williams, 1980, p.126). The assessment took fifteen man-years of work (the tenders for the first Magnox had only taken 'a hundred professional staff six weeks' (ibid.), and APC's design for an AGR was accepted. The CEGB's Chairman, Brown, called it 'a hard-headed commercial decision' (ibid, p.125).

The Decision and Organisational Interests

Despite their initial procrastination regarding the type of reactor they would prefer, the CEGB appeared to have been wholly convinced of the case for the AGR by the appraisal; their commitment to nuclear power per se became absolute and Brown even told the SCST that he would have liked an increase in the postulated programme to 8,000MW (HC 381-i, 16/3/67). Booth, a member of the Board, argued that the thirteen conventional power stations ordered in 1964 were:

> "...on the basis of our judgement at that time...the most economic ones. Since that time there has been the break-through of Dungeness B which has altered the balance and consequently our planning in 1964 has been modified on this account" (ibid).

Perhaps part of the reason for the general euphoria regarding the choice was a perceived need to outboast the Americans and their Oyster Creek results. The Board was (for once) aware of the position it played as a shop window for exports (something it is not always so keen on ((see chapter 5)) and confidently expected the rest of the world to buy British technology, forecasting that American LWRs would be overtaken by the AGR (ibid). Brown even argued that had the economic value of the offers been suitable 'we should have been very willing to go in for a water-moderated reactor' (ibid). Yet just five years later Brown's successor as Chairman, Arthur Hawkins, was to tell the SC on Science that the AGR was inherently a difficult system and 'less economically attractive than at first supposed'. Hawkins argued that the Board had been under 'fairly heavy pressure' to accept the AGR and that the reasons for the over-confidence were best explained by the AEA, as they had strongly advised the CEGB at the time (HC 444-ii, 1971-2). Doubtless, in their position as its chief adviser, they strongly advised the Government too. In the 1965 decision for the first time, with the AGR, the Board believed it had all the costings, that it could be seen in black and white that nuclear power was cheaper than other forms of power, and that the AGR was the best reactor, according to the advice received from the AEA. The CEGB was well pleased and had the assessment translated into other languages and sent to utilities everywhere. Unfortunately although there was some initial interest it became obvious that the figures were wrong and that the Board were in considerable difficulty.

By 1981 the Board were convinced that the AGR was an extravagant design and in their evidence to the Sizewell B Inquiry presented detailed costings to support their case [1]. This apparently dramatic change of heart is understandable when viewed in the context of the CEGB's underlying assumptions. Its commitment was to nuclear power after 1965, not necessarily to the AGR and it felt a lasting sense of betrayal at the treatment meted out to it by those responsible for the 1965 decision, an organisational perception that found expression in a determination to establish a professional expertise that would be equal to any in its chosen field.

This is not to argue that Brown and his senior engineers did not believe their own argument after 1965; it is patently obvious from their public utterances that they were truly convinced of the rightness of the AGR's case in the appraisal and Brown never indulged in a public condemnation of the decision-making process. This is even more important for confirming that the Board believed the appraisal's figures as, near the end of his term as Chairman, Brown and the CEGB came under considerable criticism for the delays and blunders associated with the AGR; the temptation to place the blame elsewhere must have been great. Hawkins and other senior Board members were not as ready to shoulder all the criticisms, as was shown above, but even they pointed out that the CEGB at the time of the appraisal was convinced of the breakthrough made by the AGR relative to other types of reactor. The reasons for the myopia are not clear but must be intimately connected with the advice forcefully given by the AEA and probably supported by the informal pressure of Government opinion.

Hawkins argued in 1973 that pressure had been put on the Board (see above) but he was unclear as to whether this was from the AEA of from the political system: it appears that it was probably from both. Senior ex-Board members have been clear that internal memos do not exist showing the pressure put on the Board. Burn was convinced that there was severe pressure upon the CEGB to adopt the AGR, pointing out that in 1964 it is clearly documented that Hinton was seriously considering a LWR; Hawkins's comments to the Select Committee on Science and the rumours reported in Whitehall (and given voice in the 1973 Committee by Palmer) show that the CEGB was left in no doubt by the authorities that it was expected to choose a British reactor (1978, pp.163-4). He points out that if in 1964 the utilities had been as free to choose as those in America then Britain would probably have opted for a LWR; certainly there is some truth in his assertion that it 'is hard to reconcile the hard-headed, rationalising, exploring and receptive attitudes' of 1964 with the 'AEA/CEGB consensus of 1965' (ibid). One CEGB executive appeared in little doubt, saying:

> "We were pressed to buy the AGR in 1965; but Hinton was not the sort of man to be pressed and we did get very close at that time to buying a PWR. It was all upset by APC coming in with a super design of an AGR which showed that it had the edge - on paper. With hindsight and with the knowledge that we've got today we would certainly not have accepted that tender. But twenty years ago we knew a lot less than we do now and we got taken along into that. If that hadn't happened we might have been on the water route much earlier. I think that there was pressure from the AEA via the Government, but there was also pressure from the consortia".

This view is supported by those of a member of the Commons SCST at the time who remarked that from 1962 to 1964 the CEGB was interested in LWRs and after the Oyster Creek episode would probably have bought one had it not been for the AEA dominance. The result was the decision to opt for the AGR "which wasn't even properly off the drawing-board". The deciding factors for the CEGB were, then: the APC tender; the strong and unequivocal advice of the AEA, which was the Board's official adviser on nuclear energy; and a sure knowledge that the Government wanted an AGR chosen. Because of the establishment's lack of expertise relative to the AEA and because the AEA was so forceful in its advice, the CEGB and the senior management in particular were convinced, and convinced themselves, that they had made a great breakthrough in nuclear engineering and in the costs of nuclear generation. The souring of this belief was to influence the subsequent development of CEGB R & D and the manner in which the Board approached succeeding decisions. That the CEGB did accept the AGR so enthusiastically was a testament to the influence of the AEA's experts and also marked the height of that influence. The subsequent difficulties experienced by the ESI in bringing the AGRs into production was never to be forgiven or forgotten by the industry or the political system.

It is easy with hindsight to see the AGR decision as a poor one commercially. The difficulty is in reconciling the role of the AEA as Government's unbiased and technically superlative adviser with that decision. That the Authority dominated nuclear decision-making during this period (much to the ire of Hinton) was the result of the monopoly in expertise that it possessed; indeed, in the research field it still possesses such a dominance and rejects criticism that the DEn should acquire independent scrutiny of the AEA's technical projects. A former Board member illustrated the difficulties of this by arguing that many projects are headed by Fellows of the Royal Society and asked where the Department would find a sufficient number of Nobel Prize winners to supervise them! As was shown in the last chapter, the Government regarded the Authority as the sole repository of nuclear expertise, even when it ceased to be such. Some MPs have argued argued that they had almost a free run and open access to Government at the time of the AGR decision; indeed, their constitutional position as the Government's chief nuclear adviser demanded this.

Because of this perception the Authority's advice was accepted by the Government, yet that advice was not unbiased and critical; it was proffered from a position of partisanism. The AEA's championing of the AGR was constrained not by considerations of technical excellence, but by calculations of organisational advantage. Burn claimed that it did not need hindsight to see the decision as a poor one that reflected not AEA technical excellence but the Authority's domination (1978, pp.150-163). He argued as early as 1967 (The Political Economy of Nuclear Energy) that the appraisal was biased and that it was unwise to accept the extrapolation from a 30MW prototype to a 600MW commercial reactor:

> "The AEA chose the AGR, developed it in isolation in a rather small "prototype" inadequate for extrapolation to 600MW, assessed its relative qualities and prospects very highly and promoted it strongly for use in the Dungeness B plant and as the standard reactor for the second nuclear plant programme. As principal adviser to the British Government and the CEGB on all nuclear matters, the Authority recommended the AGR as a competitive reactor promising lower costs than the BWR, and it persuaded the Government and the CEGB to make "commercial" AGRs of upwards of 600MW capacity...it was the exclusive developer and exclusive promoter of the AGR" (1978, p.152)

It is apparent that the Authority confused its role as the unbiased adviser to the system with its other role as the developer of reactor systems and this was one of the main reasons why the Select Committee on Energy was to castigate the Authority for giving poor advice and recommended that it restrict its role to scientific research and development, leaving reactor development to those intimately concerned with its commercial applications (HC 114-i, para. 71).

The Authority had been established with limited objectives, its primary function being to develop the atomic bomb for Britain and then to explore the possibility of civil exploitation. It had carried through these goals with such commitment and with such conspicuous success that the teams responsible had followed through the achievements with a determination to develop the full potential of 'their' research. The professional attachments analysed in the last chapter combined with the technocrat-dominated hierarchy of the AEA to secure a commitment to the goals of the teams. That is, due to the technocratic style of decision-making within the Authority and the tradition of technical precedence, even in the more commercially-minded Production Group, it was inevitable that the Authority would come to regard the goals of its research teams as its goals within the reactor field (a situation that did not prevail in other areas such as fuel fabrication and fusion research, see below). The Authority, after it had successfully developed atomic weapons, was identified by the political systems in terms of its reactors; therefore it felt that it had to secure the adoption of its reactor: a threat to that adoption was a threat to the Authority itself.

On this a former AEA Board member was quite candid, arguing that if the AEA had 'lost' that decision then it would have been a 'blow to the Authority's morale'. This perception of the appraisal in terms of a competition is particularly revealing in the light of the Authority's 'unbiased' role vis a vis the ESI and Government. The confidence the Authority had that its advice would prevail was underlined at least a full year before the appraisal when it revealed plans to moderate the Capenhurst diffusion plant to supply low enriched fuel explicitly for the envisaged AGR programme. The development at Capenhurst was a major and expensive project and would not, could not, have been undertaken on a mere off-chance (AEA Annual Report, 1964-65, pp.10-11). The Capenhurst development can also be seen in the light of the analysis of Chapter 8 and the evolution of the Production Group's commercial ethos, leading eventually to the establishment of BNFL. The cutbacks in the teams of the Production Group following the 1960 White Paper reducing the atomic programme, due to severe CEGB pressure, had alerted the Authority to the consequences of a reduction in the commitment to the expansion of nuclear power. A decision to adopt the American LWR would have caused similar unemployment amongst Authority personnel and would have threatened the AEA itself; little wonder that the AEA lobbied so furiously and in contrast to its brief of impartiality.

The politicians were similarly trapped by organisational constraints. Just as the narrow front approach and the decision to develop exclusively gas reactors had resulted in the AGR and this in turn had become the AEA's flagship after the delays in the development of the Fast reactor, so the political tradition of exclusively relying upon the Authority for its technical advice constrained the Government. Effectively, the AEA's flagship project became the Government's. A senior back-bench MP argued that because of the prestige and importance given by the Government to nuclear power after the Suez crisis and because of the technical dominance of the AEA, the policies of the Authority were adopted by the Government, providing that they accorded with the Government's broad political strategy. Because the Government had invested such large amounts of money and other resources into nuclear energy (sums impossible to calculate (Benn, 1984)), and more importantly into AEA-led and inspired policies, it would have been difficult for the

Government at any time to act against AEA advice since that would have meant questioning their entire policy; there was not the political will to embark upon such a fundamental reappraisal or the confidence to do it.

Until the full extent of the AGR's problems became painfully apparent the DEn, which advised ministers (and its predecessors MINTECH and the Ministries of Fuel and Power) secured its nuclear advice almost exclusively from the AEA. Thus political considerations were taken against a background of technical advice that was seriously biased; indeed, the calculations regarding the costs and potential of the AGR were nothing short of wishful thinking until about 1970 when a semblance of reality was forced upon the policy system via the realisation of horrendous cost overruns, delays and construction errors that beggared belief. It was a realisation of the paucity of technical opinion involved in the policy process that led both the Commons SCST in 1967 and the Select Committee on Energy in 1980 to recommend that the AEA concentrate on its primary task of R & D and in 1980 to argue that the DEn should secure alternative sources of advice and seek to monitor the technical projects of the AEA and the advice tendered from the organisation. As noted above (Chapter Four) although the Government said in its Green Paper reply that such a duplication would be a wasteful use of resources, it had by 1980 supplanted the AEA with the CEGB and business interests as its main adviser in the field of civil nuclear energy.

The views of the politicians were shaped by nationalism and the AEA and its own brand of parochialism, a parochialism that Hill later disavowed. A senior CEGB executive argued that from the early 1960s there was a growing realisation within the AEA that they were fighting 'a rearguard action for their continued existence' but that no one had the political courage to order the termination of the Authority. This realisation contributed to the Authority's determination to rubbish foreign competition, a determination another CEGB Board member said was 'sad' because 'some engineers just wanted to cling on to British technology' because 'there is a certain pride in saying something is invented here'. He argued that 'one of the dangers of our engineers is (that) they cling onto a British-invented product on the basis that it's "better" than anything anybody else has invented. A view that easily transmitted itself to politicians eager to increase British exports and jobs. Fred Lee, the Minister of Power, called the announcement of the AGR's success in the appraisal 'the greatest breakthrough of all time', something Burn understandably calls 'among the most absurd of ministerial statements' (1978, p.10). Yet Lee was not alone in his public euphoria at the British 'breakthrough'. In the Lords, Lord Byers sought to congratulate the AEA for 'a really remarkable technical breakthrough' and Lord Sherfield (who as Sir Roger Makins was a former Chairman of the AEA) said that 'this decision will indeed be very welcome to the scientists and engineers of the AEA, who have developed this system' (Atom, no.10-4, June 1965, pp.99-101). Whilst Frank Cousins, as the Minister of Technology, wrote to Penney that it was 'particularly gratifying' that on 'sheer technical and economic merit, the AGR has shown itself more than capable of holding its own with its competitors' (ibid).

The AEA's claim that the AGR would produce electricity at least ten per cent cheaper than the most modern coal-fired station (ibid, p.168) led other observers to claim that British engineering had scored a 'notable triumph' by a 'most convincing margin' and that the all-important exports were badly needed (New Scientist, vol.28, October 1965, p.179). There was a decidedly bipartisan approach to nuclear power, there was no such thing as a Conservative or a Labour reactor. Certainly the expansion of nuclear power accorded with the main policy goals of both major parties in that it provided a source of energy that was not strategically vulnerable to Middle Eastern producers and it helped to close a gap in expected demand and actual coal production. Also it pandered to expressions of national pride, promised exports and helped to create jobs, particularly in the depressed North East. Both Conservative Governments, committed to private enterprise within a Keynesian framework, and the essentially Keynesian Wilson Government, committed to a technological revolution and export-led economic expansion, were enamoured with the prospect of nuclear expansion. The AEA in promising to fulfil the above expectations with British technology were able to influence Government decisions on the matter (Stewart, 1977). In short, the AEA told the Government that the AGR was what it was looking for and the Government joyfully acceded to the charade.

So committed to its overall technological aims was the Government and so skilful was the AEA at presenting its policies as being in accord with those aims, that the costings of the AGR appraisal found their way into both the First and Second Fuel Policy White Papers (1965, cmnd.2798 and 1967, cmnd.3438). Crossman wrote that the Second White Paper was totally written by officials strongly influenced by the AEA and he wondered at the phenomenon of Ministers of Fuel and Power with working-class mining backgrounds being seduced by the arguments of officials and opting for nuclear power at the expense of coal and miners' jobs (in Pearson, 1981, pp.169-189). Both Papers envisaged a growth in nuclear power, with the second one making a strong assumption of cheap nuclear stations grabbing a larger share of the base load for electricity generation. This was in full accord with the Government's aim of supplying cheap electricity to consumers so that British industry would have lower generating costs than foreign rivals and would therefore gain a competitive edge in export markets (ibid). The 1967 paper expected that most future power stations would be nuclear. The importance of AEA influence is illustrated when it is realised that nuclear costings in the Government's projections based upon those given in the Dungeness B appraisal were, within five years, to become a laughing-stock.

At least one former CEGB Chairman believed that Britain opted for the AGR because it was the safe political choice, as well as appearing to be superior on the drawing board. As was noted above, at the time of the appraisal there was no clear world reactor: even the Americans had not plumped for the PWR or the BWR as their leading reactor; the Oyster Creek contract had gone to a BWR, but the PWR was eventually to secure the majority of orders. Britain had the Magnox, the Canadians the CANDU, the French and Germans had not decided to standardise yet, 'therefore the safest political bet was the AGR', which if not better than the others could not be demonstrated to be worse. In addition, it provided jobs and profits for Britain as well as being a public vote of confidence in British research and development and the massive investments made by Governments in that R & D.

This view is fully supported by a civil service Under-Secretary concerned with nuclear energy for many years until the announcement of the adoption of the SGHWR. He argued that there was no documented evidence within Whitehall to support the belief official pressure was put upon the CEGB to adopt the AGR, which surprised him as there was a tremendous amount of vested interest in the choice of the AGR. He felt that the pressure would probably have become official had the strongest contender, the LWR, been chosen. He too believed that there were organisational constraints that led inevitably to the choices made, arguing forcefully that from his experience of dealing with them when the technocrats had devoted the greater part of their working life to a project and their career was based upon it, then they were going to identify with and become attached to that project, no matter 'what rational justification they make for their viewpoints: the attachment is an emotional, a subjective one' and the same 'could be said for an attachment to an organisation'. He felt that at the time of the Dungeness B decision the AEA were committed to the AGR and lacked the objectivity their legal position called for: understandably so as a LWR decision would have diminished the role and prestige of the Authority.

Thus even though they attempted to justify themselves rationally, the advice was political not scientific and technical; it was geared to what the Government wanted to hear and in turn influenced the politicians. If not a conspiracy, there was certainly a remarkable accord between what the Government and the AEA wanted and the decision the ESI eventually made. As is argued above, it can be seen that formal pressure does not need to be put upon members of the Board of a nationalised industry; there is a wealth of informal measures that are equally efficacious and when these were combined with the technical arguments put by the AEA the choice was decided.

The Lessons of the AGR Decision

The lessons of the AGR decision fall into two groups: lessons for the decision-makers; and, second, lessons for the study of professional influence in decision-making. For the decision-makers the strange role of APC as the successful consortium and the subsequent history of the AGR programme led to a commitment to reform the consortia structure and to review the relationship of the private and public sectors.

It can be argued that APC provided a way for the AEA to secure the adoption of the AGR, and the evidence presented to the SCST in 1967 suggests a degree of collusion not repeated elsewhere in the relations between public and private organisations. The design of

APC's AGR also represented a way in which the ESI and the Government could choose and sanction an AGR in the face of stiff American competition without losing credibility (at least initially) because the figures in the design were so impressive. Both Burn (1978) and Williams (1980) go to considerable lengths to illustrate the defects in the design of the APC tender; indeed the impression they present is that it was little more than an artist's sketch, yet it was submitted after considerable AEA/APC collusion - indeed it was effectively an AEA design. This degree of co-operation was denied to the other consortia. NDCL told the SCST that they were completely tied to the AEA's research and although they would have liked water research to continue after 1958, they could do no more than 'make representations' (HC381-i, 6/4/67). TNPG repeated the desire to have recourse to alternative reactors. Sir Edwin McAlpine, the head of the company, argued that the AEA designed and built all of the prototype reactors and although 'they do an excellent job' they did not do it in a commercial atmosphere; they did not design products to (meet a customer's rigid specifications with a firm price' (op.cit. 13/4/67). Both TNPG and NDCL argued that they had close co-operation with the AEA and although NDCL wanted even closer liaison over design, neither group felt that the close and detailed help given to APC by the Authority was unfair. Colonel Raby, the Chairman of APC, vehemently denied the suspicion of unfair advantage, although as Burn points out he did have a special personal relationship with some of the AEA staff, having been Cockcroft's assistant at Harwell before joining APC (1978, p.164). Yet TNPG were not allowed to deviate from the design specifications published by the CEGB, whilst APC were not only allowed to deviate, but were actively aided in doing so by the Authority. Raby argued that:

"I am pretty certain that this would also have been put to the other two consortia. I have never known a case yet where the AEA have given any sort of information to one consortium without giving it to the others". (ibid, 20/4.67).

Moore, a member of the AEA's Board denied the charge of one member of the SCST, Sir Harry Legge-Bourke, that there was a 'special relationship between APC and the Authority, and that it appeared that 'immense pressure' was put upon the CEGB to allow a change of design specifications in order that APC might submit their 'impressive' design (ibid. 7/6/67). Burn is in little doubt that APC acted to present an AEA design and secure the adoption of the AGR and that in the process it was given special help (1978, p.120-122, 164-167, 170, 178, 197), he argued that this was felt necessary in order to meet the new threats in fuel efficiencies perceived to come from LWRs as well as the usual desire to ensure the defeat of the American challenge. The public affirmations of TNPG and NDCL on the even-handedness of the Authority were not, he argued, repeated in private, the reason for the public statements of faith being that the consortia were dealing with two public monopolies and needed to secure future orders.

That there was some considerable measure of collusion between the AEA and APC there can be little doubt, for the AEA design teams worked closely with all the consortia. To what extent this collusion was greater than that with TNPG and NDCL cannot be verified due to the lack of published (and indeed unpublished) material. It is known that APC benefited from AEA help to the extent that the APC design was effectively an AEA design and that in presenting it APC were able to depart from CEGB specifications to a greater extent than the other consortia; however, this may just have been because APC was the weakest consortium and both the AEA and CEGB were trying to keep it alive. The flaw in this argument is that the Authority were already pressing for the rationalisation of the industry. On balance, the assertion that APC was a vehicle for the AEA's design proposals appears a convincing argument. It would be in keeping with the organisational goals of the Authority to perform such a manoeuvre. The important lesson for the decision-makers was that the weakest consortium achieved the contract with the worst design; this was followed, not by another competition, but a return to Buggins's turn and TNPG being awarded the next AGR and so. Despite Brown's vigorous claim that:

> "I do not recognise Buggins; I do not recognise him at all...we are strongly persuaded of the benefits of competition" (HC 381-i. 4/5/67),

there was no competition after Dungeness B; the CEGB simply awarded the contracts. This was a curious way to behave in that Brown stiffly resisted the Authority's and the consortium's attempts at rationalisation, arguing the benefits of competition, and then avoided the logical practice of that belief. It can be argued that the CEGB preferred to retain the potential of competition to frighten the consortia into behaving; in the event the inherent weaknesses of the system led to the collapse of APC and the need for the CEGB to buy them out. The weakness of the design also meant that the station had to be completely redesigned and practically rebuilt; as much of the inside had to be ripped out and backfilled. The cost overruns and delays made Britain a laughing-stock in the nuclear field and probably did much to ensure that no export orders were forthcoming, America sweeping the international market. The folly was compounded in that of the five stations ordered (four for the CEGB and one for the SSEB) there were three different designs, 'all of which required considerable extrapolation in engineering terms from the 30MW Windscale prototype' (CEGB, 1982, p.17). The lessons were obvious: the consortia had to be reformed and the designs had to be standardised. The CEGB learnt the lesson well and future AGR orders were indeed standardised.

It was not that the decision to build a British reactor was wrong, just that the AGR was not ready. The veracity of the British policy-makers would have achieved more credibility with foreign buyers had they simply said that they did not know which was the better reactor and were therefore going to buy British because of the need to support British technology and industry. The facade of rational technological decision-making laid the system open to charges of technical ineptitude, particularly when the failures became apparent. It was clear from early on in the programme that there was a need to rationalise the industry and standardise around one design, but Brown was resolutely opposed to both, arguing the need for competition and that "nuclear power is so young and rapidly developing...it is a denial of development to replicate the designs," or as he called it to embark on 'Chinese copies' (HC 381-i, 16/3/67).

This is clearly based upon his experience in the 'Brown revolution' in conventional stations (see last chapter), but it came to be seen as wrong. Yet perhaps he was judged too harshly by his contemporaries and successors in that to have standardised around the Dungeness B design would have had frightening consequences, so premature was the decision to build it and so incompetent the design. By 1970 the ESI was unanimous on the need to reform the industry and to standardise; the CEGB however wanted to standardise around the PWR, having been thoroughly put off the Authority's AGR and the method of building it. Board engineers and execitives, amongst others, described the engineering as watchmaking by the ton on site and they felt the system contributed not only to delays and cost overruns because of design difficulties but also, because a large proportion was machined on site, it contributed to union power and made the programme particularly vulnerable to labour militancy. This is returned to below and in the following chapter.

If the ESI learnt the lessons of the AGR decision, the politicians and the AEA did not. The Dungeness B appraisal had shown the weakness of political decisions based upon limited technical advice when that advice was crucial. The decision was not a technical decision; it was political, yet it required a degree of technical input which the AEA was responsible for providing. In recognising the political requirements of the policy-arbiters the Authority was quick to put to its own advantage the technical ignorance of the politicians and civil service and their readiness to trust to the advice of the AEA staff. The biased and fundamentally inadequate nature of that advice was exposed by the AGR decision; yet in the instance of the SGHWR decision the politicians (though not all the civil servants) were to repeat the mistake of relying upon AEA advice to the exclusion of the other parts of the nuclear policy-system. The AEA should have learned the folly of extrapolating a full-sized commercial reactor from a small prototype; it did not. Instead it compounded the folly with the SGHWR.

In illustrating the political rather than the technical nature of the decision-making process, the AGR decision also showed the role of technocrats in that process. Rooted within their employing organisational hierarchies, the technocrats sought to realise professional and organisational goals through slanting the advice they provided to the politicians to accord with the goals of those politicians. They were influential when they told the politicians what they wanted to hear; thus the presentation of technical advice determines its effect upon the policy-making process as does the esteem in which the organisation providing the information is held in Whitehall. The high regard in which the AEA was held at the time of the AGR appraisal led to greater weight being placed upon its advice than that of the CEGB, added to which the AEA adduced the two Magnox exports as evidence that it knew how to design machines which would ensure exports for the manufacturing sector. Commercial success for the nation being a goal of governments of whatever political colour in a liberal democracy, it follows that technocrats able to exploit that goal to their own advantage (either through sharing it or appearing to share it) will have a greater influence than those who appear to have a contrasting goal.

The CEGB's interest in American technology appeared almost like an act of betrayal to the AEA and the Government, intent as they were on a desire to boost British R & D and British exports. To the Government, support of the AEA seemed the best way of achieving these goals and the AEA, with its own organisational goals of survival through expansion, did nothing to dissuade the Government of this notion - a theme that recurs repeatedly throughout this book as technocrats are seen to ally themselves with the political system in order to achieve occupational goals and influence policy. It is not confined to Britain; Dixon (1973) and Price (1965, pp.86-134) note similar behaviour in the USA. The nuclear policy-making process was not a professionalised policy-system per se, but it was dependent for much of its advice and for all of its options upon professionalised organisations or sub-systems. The nature of the technocratic options led to their being adopted in that they appeared to the Government to advance political aims; the choice for the Government was to choose the technocratic options which most accorded with those aims. In the example of the AGR decision, it was the AEA that predicted political goals correctly and rigged its advice in anticipation of this. The influence of the scientists and engineers in the choice of the AGR was therefore profound. But it was not the advice of all technocrats, only of those engaged in gas research within the AEA, and their allies in the upper tiers of the management structure. That influence, although profound, was also heavily contingent upon its political acceptability. These political lessons were an invaluable part of the political education of the technocrats in the CEGB and elsewhere, particularly when the AGR programme fell behind schedule and overran costs. From 1970 the Board began to exhibit a growing belligerence in its dealings with the AEA and confidence in dealing with politicians. The effects of this were seen during the SGHWR and Sizewell decisions, to which the following chapter addresses itself.

NOTES

1. See 'Thermal Reactor STrategy', 1981. And volumes of CEGB's statement of case fo the Sizewell B Public Inquiry, especially Vol.I.

7 The reactor decisions II

The Introduction to the previous chapter explained how it should be viewed as the concomitant to this, the two chapters together comprising the shift in analysis from a more rigid investigation of professional structures, to an exploration of organisational and professional motivations. The sequence of this chapter takes up the series of reactor choices where Chapter 6 left off to probe the SGHWR and PWR decisions of the early and late 1970s respectively, a sequence that runs over into the Sizewell Inquiry of the 1980s. The sections of this chapter progress from the previous analysis of the AGR decision to examine the SGHWR decision from the perspective of the organised pro-SGHWR and pro-LWR lobbies, lobbies that to begin with divided along the established lines of the AGR debate. The chronology of the debate then leads to a consideration of the choice of the PWR made by the Conservative Government in 1987 and a discussion of the rationale for the choice which includes the argument that it represented the failure of nationalism whilst coinciding with the market orientation of the Government. As with the previous chapter, although the sequence is chronologically determined, the analysis eschews a strictly chronological approach, concentrating upon the organisational, political and professional motivations that provided the dynamic for change.

The severe technical problems associated with the AGR had combined by the mid-1970s with a vestigial opposition in political circles to American technology. The Labour Government of 1974-79, therefore, decided to equip the Generating Boards with the SGHWR. It soon became clear that the costs of developing this reactor were unbearable so two more AGRs were ordered as a stop-gap and work began on them. The Conservative Government of 1979 was imbued with the principles of commercial enterprise and independence, goals exploited by the pro-PWR lobby to secure Government backing for that product. In charting the twists and turns of the policy process, this chapter aims to provide an explanation for the choices eventually made; the arguments propounded stemming from the perspective of the analysis of technocratic motivation that has been adopted.

The SGHWR

If the decision to adopt the AGR represented the peak of the AEA's influence, then the decision to select the Steam Generating Heavy Water Reactor was the point at which the watershed began to turn. The choice of the SGHWR by Varley in July 1974 was a vistory for the AEA that within just two years could be seen to have been Pyrrhic, leading to a substantial loss of confidence in the Authority by the political system (Ministers and civil servants) and a corresponding fall in the ability of the Authority to influence the decision-making process. Indeed, the decision to adopt the SGHWR almost guaranteed that the option to pursue the PWR would be granted to the CEGB, as it vindicated those within the ESI who had consistently opposed British technology and turned the Whitehall machinery solidly pro-PWR, although some senior civil servants had been in favour of LWRs from the late 1960s when the troubles with the AGR first became apparent.

It is argued below that Varley's decision to adopt the SGHWR was completely political, for it had very little to do with technical suitability and much to do with the political fortunes of the Labour Government, dependent upon the trade unions' co-operation for its survival. Whereas the AGR could be seen as a system adopted before its time, or at least before it was fully developed (it is worth noting that from 1980 the AGRs performed impressively, after the design faults had been solved) the SGHWR was a system adopted after the period when it could usefully have been developed. That is, if Britain were going to fruitfully pursue that option then the SGHWR should have been adopted and developed instead of the AGR, as it was the narrow front approach ensured that the AGR received the bulk of funding whilst the SGHWR was kept ticking over at Winfrith Heath under first Cook and then Fry. It could be argued that the decision to continue with water development at all represented a substantial victory for those teams in the Authority who were not convinced of the wisdom of concentrating entirely upon gas; certainly there were severe disagreements over that decision and many of the water advocates left the Authority [1], although one, Lord Marshall, eventually came to be Chairman.

Symes points out (in Marshall, 1983, pp.334-344) that the heavy water reactor development underwent its first serious study in 1956, just before the decision to concentrate upon the AGR; and that from about 1957 the Authority decided to fund its development (albeit at a much lower level than the AGR) in order to ensure there was an alternative available should any insurmountable difficulties emerge with the AGR; it was the AEA's insurance policy. When the period of improved confidence in the AGR occurred in the early 1960s the SGHWR was under some threat; but the advances claimed by the Americans and

Canadians convinced the AEA of the need to remain abreast of water technology (Symes, 1983, pp.344-346). It is somewhat ironic, considering the enormous advances made by the Canadians with their CANDU, that both SGHWR and CANDU spring from the same wartime project at Chalk River where Britain and Canada collaborated in early military and commercial nuclear research. When Varley came to make the decision for the third nuclear programme, the failure of the AGR meant that there was only the SGHWR available as the British alternative to foreign reactors. The AEA duly dusted-off its prototype at Winfrith Heath and presented it to the Minister as a sound option for the CEGB to buy, and Varley believed them. This is not to criticise him: as a career politician he had few alternatives and a limited technical knowledge. It is to point to the weakness of basing decisions upon a narrow partisan source of advice.

Apart from the engineers involved in the development of the SGHWR, few people believed it to be a reactor capable of extrapolation to full commercial operation in the UK. It was viewed as an intermediate reactor more suitable for export to countries seeking a small nuclear power station. Penney in 1967 expressed severe doubts about the SGHWR's future and said that if it had a future then it was an export earner to countries seeking a station in the 300MW range. It was not suitable, he thought, for the CEGB who wanted stations of at least 1000MW; indeed perhaps only the SSEB and the Northern Ireland Board in Britain would want it, although he believed even this to be doubtful (SCST HC 381-i, 2/3/67). By 1967 the consortia felt that the SGHWR was already passe. McNeil argued that:

> "I have the feeling that (the SGWR) is a little late in the race because it is a water reactor and the others have been well established for a number of years, so it has a lot to overcome." (ibid, 6/4/67).

Although TNPG worked closely with the AEA and even had a senior design engineer at Winfrith, they were not enthusiastic regarding its commercial potential, pointing out that their teams were hired on a contract basis and the full responsibility for development lay with the Authority (ibid, 3/4/67). Benn, as Minister of Technology at the time of the AGR decision, saw the SGHWR in terms of his AEA brief, arguing that he thought it would meet the needs of countries with a smaller electricity demand; but he left such considerations to the AEA who developed all Britain's prototype reactors 'entirely upon a scientific assessment of what systems were worthy of further study' (ibid, 11/5/67). It is interesting that the concern for commercial exploitation was not one the Minister associated with the AEA. The SCST were however anxious that the SGHWR be considered as a future reactor system for the ESI and recommended that work on its development be speeded up (ibid, para.110). By 1973 the MPs' opinion had been even more swayed towards the SGHWR (HC 145, 1973-4), but this was partly due to the intervention of Sir Francis Tombs, Chairman of SSEB and an advocate of the SGHWR. Tombs's view accorded closely with that of Symes, a power engineer who worked on developing the SGHWR and who wrote:

> "The claims made for the SGHWR system...all seem to have been endorsed by the building of the prototype and by its operation...there would seem to be no basic technical or

economic reason which should, in principle, have prevented the reactor from taking its place in the electricity generating industry, for which it had been specifically developed". (1983, p.366)

Tombs argued that the reason the reactor did not take its place in the generating industry was because the CEGB killed it with kindness', a view examined below.

Because of the evident failure of the AGR to fulfil the promises made for it, by 1972 when the time came for a third nuclear programme to be selected, the AEA realised that it had to produce an alternative to the LWRs or the CEGB would select the American system and as was the case with the AGR decision, such a selection would inevitably spell the demise of the AEA's technical dominance. In 1973-4, the Conservative Government became aware of the costly blunder in AGR construction and was moving hesitantly towards the CEGB's position, it was a strong probability that a LWR would have been selected had Heath not lost office in 1974 and Labour thrown the situation back into a state of flux. Aware of this the AEA immediately resumed lobbying for the SGHWR, concentrating upon the fact that it was a British reactor, creating jobs in the UK and more importantly providing industry with desperately required orders. As Benn told the Select Committee on Energy in 1980, the need to maintain the manufacturing capability to construct large power stations in Britain demanded that there be a continuous process of ordering; nuclear orders had slowed down dramatically and both he and his predecessor were concerned with protecting the industry and the tens of thousands of jobs therein; the alternative was a massive import bill (HC 114-II, pp.420-428).

To aid the industry in its choice of reactor the Government had established the Nuclear Power Advisory Board in August 1972, and had set them to work on thermal reactor policy in September 1973. The Board was chaired by the Minister and consisted of the Chairmen of the AEA, NNC, CEGB, SSEB, the area boards and the Electricity Council: its report was published as a White Paper in September 1974 but did not really resolve the question of reactor choice, the NPAB dividing internally along organisational and therefore SGHWR/LWR lines (Williams, 1980, pp.227-234; and cmnd.5731). The NNC and the CEGB joined forces to oppose the SGHWR, the NNC's members arguing that the only type of reactor they had any hope of exporting was a LWR. Burn argued that the decision to adopt the SGHWR simply compounded the problems of the AGR decision and was made by a Minister as an assertion of political power: it was a choice by a technically ignorant man between two sets of experts (1978, pp.278-300). Anxious to be seen to be supporting British technology and defending British jobs, his choice was understandable, and the AEA provided him with the necessary technical rationale. Burn's harsh judgement therefore missed the real political constraints which channeled Varley's decision.

The pro-SGHWR Lobby

The prototype SGHWR came onto full power in January 1968 and was officially opened by the Duke Edinburgh the following month (AEA Annual Report, 1967/68). The AEA first publicly referred to the SGHWR as something more than a 'complementary' system (Annual Report, 1968/69, p.4) in a 1970 review (Annual Report, 1970-71), and by 1972 was devoting increased resources to what it now called a 'viable alternative to he AGR' (ibid. 1972/3, p.3). The reasons for the AEA's advocacy are clear, but it acquired new allies within the trade union movement, particularly those concerned with heavy engineering. The TUC's Energy Committee in making representations to Varley said that although the choice was difficult, it preferred the SGHWR (EPEA Annual Report, 1974). This advocacy of the SGHWR was in accordance with the tradition of support that nuclear power had enjoyed form its earliest days within the trade union movement. Any expression of opinion from within the unions must have been accorded additional emphasis by the Labour Government of 1974-78, particularly in between the two elections of 1974 when a tenuous Labour lead could have evaporated in the face of the type of trade-union militancy that caused Heath to call the February election, as Stewart (1977) and Ashford (1981, chapters 3,4) show in their studies of the period. Consequently, in making his decision, Varley would have weighted the unions' preferences heavily, and it appears that he certainly weighted them heavier than those of Hawkins and the CEGB. The resolute opposition of the CEGB to the SGHWR was overruled by the Minister (see below).

The unions linked their desire for the SGHWR to a belief that the LWRs were inherently less safe, a belief fostered by the AEA from the early 1960s. Such questions of safety had once held sway with the CEGB; Brown told the SCST in 1967 that he had some doubts about LWR safety (HC 381-i, 16/3/67). Part of the AEA reserve on the subject had come from the strict parameters in which they had been forced to work, aware that they were designing systems to operate in the densely-populated British Isles and lacking the luxury of America's vast wilderness. In 1972-74 the Government's Chief Scientist, Sir Alan Cottrell expressed some concern regarding the reliability of the PWR's safety vessel, a concern he was to express again to the SCE in 1980 (Williams, 1980, p.224, for 1973, and HC 114-i, p.120-124, for 1980). Pressure vessel integrity was to be a continuing subject of debate between the various experts, the response time available to react to an accident in a PWR was considerably less than in an AGR. Both Hinton and Rotherham told the Science Committee in 1973 that in their opinion the LWR was inherently less safe than the SGHWR or gas reactors (Williams, 1980, p.225), opinions that, due to the stature of the proponents, carried considerable weight. As Williams points out, to be fair to them the industrialists and the CEGB were not immune to the safety arguments, but pointed out that since LWRs were the world's most popular reactor there had been considerably more research done on the safety aspects of them than other types of reactor and any outstanding doubts were not insurmountable, indeed, because of their popularity it was likely that such problems were more likely to be solved than those infesting British reactors because the world-wide research effort was so much greater (1978, p.226-7).

The most telling intervention in the debate came from Sir Francis Tombs, Chairman of the SSEB and also a member of the NPAB. His evidence to the Select Committee and to Varley, when combined with the Chief Scientist's and the trade unions', swung the decision for the SGHWR. During the SCST's investigations of 1973/4 (HC 145, evidence HC 73) Tombs argued strongly for the SGHWR: An MP recalled that he came before the committee and 'was very convincing', earning himself 'a lot of political support'. The MP continued that it was not just what Tombs was saying that was important, but also his presentation. He was passionately devoted to the SGHWR and spoke to the MPs in their 'own language', whereas Hawkins always appeared with a bank of advisers. This reflected the cautious, almost pedantic, approach of Hawkins to his duties, supporting his arguments with a wealth of data. The clash between the CEGB, the AEA and SSEB reflected the disjointed nature of the engineering occupations, each group of engineers arguing from an organisational, rather than a professional standpoint, which tends to confirm the arguments of Chapter Three above regarding this aspect of the technical professions. It is in some contrast to the united front expressed by the older social professions, for example, doctors when a reform of the Health Service is threatened (Dingwall and Lewis, 1983, pp.84-106).

The divisions within the ranks of the technical experts had by 1973 become so severe that they made a considerable impression upon the politicians. A member of the SCST recalled the confusion of MPs faced with the advice of highly-skilled and respected engineers who fundamentally disagreed with each other. This division made the eventual decision of Varley appear even more overtly political, forcing him to choose between his own advisers on a technical question that he was manifestly technically unfit to answer. The desire of the politicians to have a form of consensus within the technical ranks is to spare the Minister that type of problem and was the reason why the Powell and Vinter committees and the NPAB were established, consensus among experts for the minister's decision allows that decision to be presented as a technical inevitability: it depoliticises the issue and lends overwhelming support to the course of action pursued. Conversely, technocratic squabbling leads to contention and political uncertainty which, because they are technically ignorant and therefore require guidance, the politicians abhor. A former minister was clear that at times he wished the policy-system had been less flexible and a little more rigid as it would have made his job easier, as it was, he always ensured that when he spoke with a scientist he knew whether the technocrat was pro-LWR or anti. The Tombs intervention and disagreement with the CEGB created just the sort of dissensus the politicians sought to avoid; but it also provided them with a way to opt for the British SGHWR and keep the unions quiescent. Because Tombs represented the second-largest organisation within the ESI, the politicians could argue that they were not opposing the ESI per se, but only part of it. The Tombs-Hawkins split ameliorated the option of the SGHWR and also helped prevent the adoption of LWRs for another decade.

Tombs's arguments were well rehearsed and convincing, and no doubt they helped mark him down for higher office. He said that electricity demand would not grow as fast as the CEGB predicted (he was proved correct); that the Board were over-reacting to the problems of the AGR, which were essentially those of poor organisation, not engineering; and that the PWR was not available 'off the shelf' as the CEGB argued but was a reactor that had a number of safety questions connected with it that were not connected with the SGHWR. Principal among his objections to the PWR were questions relating to the integrity of the pressure vessel and the emergency core cooling system, problems which could lead to a loss of cooling accident (LOCUS) similar to that which afflicted the station at Three Mile Island. He said, therefore, that he would not be prepared to support the building of the PWR for a number of reasons which were technically based. If the UK needed nuclear power quickly, though he did not think we needed it as quickly as the CEGB, then the SGHWR offered a very good argument.

The reasons were that the 100MW reactor at Winfrith Heath had been running for some years then, and running extremely well, and it was modular in construction. It was intended to be a 600MW station: the fuel clusters, the channels were exactly the same, so it wasn't going to be a very big extrapolation. The analogy to the SGHWR is the CANDU. The difference is that CANDU is cooled as well as moderated by heavy water, but the SGHWR uses heavy water in a static form, it is an enclosed system and therefore less expensive and better environmentally than the CANDU

He also pointed out that because it was static there was no need to build heavy-water plant, and that although it used enriched fuel, so did the AGR and the UK had excellent enrichment facilities. He also argued that much of its construction was similar to the CANDU and they were the most reliable reactors in the world in his opinion. There are some grounds for believing that in addition to the arguments of the trade unions and the needs of manufacturers (who mostly wanted LWRs) Varley saw Tombs as a casting vote in the struggle between the AEA and the CEGB. As was noted above, the AEA as a whole saw the SGHWR as another chance to keep American reactors at bay.

It must be noted, however, that even though the AEA promoted the SGHWR, Hill at least was beginning to have second thoughts, although these were not expressed publicly. He was concerned that the UK was being isolated from the rest of the world technologically and that attention was being focused upon the thermal reactors at the expense of the FBR. In the NPAB his advice began to sound a little ambivalent, and privately it is probable that he already foresaw difficulties with the system and felt that the UK would have been better off on the LWR route. In the event these fears did not surface at the time and Varley was swayed by the evidence and political considerations to adopt the SGHWR.

The LWR Lobby

The CEGB and most of the manufacturers were in favour of adopting the PWR, as was the DEn's chief scientist, Walter Marshall, who was also a member of the Board of the AEA (a position illustrating the close DEn/AEA links, although his advocacy of water reactors was not in line with standard AEA positions). It was probable that he helped convince Hill of the argument for LWRs: in any event his views led to Benn replacing him as Chief Scientist with Bondi, and then to Mrs. Thatcher's Government appointing him first to the Chairmanship of the AEA and then the Chairmanship of the CEGB (and a seat in the House of Lords). Because of the delays and expense of the AGR programme, the CEGB was opposed to any more gas reactors and they viewed the SGHWR as another undeveloped prototype that was not ready for full commercial production. Their experience of AEA prototypes had convinced them to seek to avoid them and they vehemently argued against it, calling for a PWR option because it was tried and tested and they could simply go out and order as many stations as they required, knowing exactly what they were going to get each time. The CEGB rested their case for buying the PWR on three grounds:

1 that the UK required a large tranche of PWRs very quickly, something like eighteen 1000MW stations over the following decade;

2 that gas technology was obsolete and the AGRs were languishing late and over priced;

3 that the PWR was available 'on the shelf' and was therefore ideally suited to a rapid programme.

Part of the reason for the CEGB's perceived need for a massive increase in its nuclear capacity were the effects of the miners' strikes in 1972. As was shown in the last chapter, upon taking office as Chairman in 1972, Hawkins was of the opinion that there was a need for a reduced ordering load; indeed, that for the foreseeable future the CEGB needed no new nuclear stations at all. The miners' strikes of the early 1970s quickly followed by the OPEC price rises after the Middle East wars of 1973, exposed in stark and brutal detail the strategic vulnerability of coal and oil to interruptions of supply (CEGB Annual Report, 1971, pp.1-22; 1973-4, pp.7-11, 1974-75).

The preferred reactor, for the reasons outlined above, was British-built PWR technology, licensed from Westinghouse of America. In explaining his preferences to the SCST, Hawkins was asked by Palmer whether he wanted nuclear power per se, or LWRs, although Hawkins replied that he wanted nuclear per se, with a preference for LWRs, according to opne MP the Committee was left in no doubt that his aim was a large tranche of LWRs. To a certain extent, the antipathy to the SGHWR and the gas reactors had grown since the appointment of Hawkins who developed a single-minded approach. As late as 1969, Brown was still considering the High Temperature Reactor which the AEA was developing and even believed that the SGHWR could be commissioned as a complementary system to the AGR in the mid-1970s (Williams, 1978, pp.210-202). By 1970, this was changing and when Hawkins did his own reassessment following the miners' strikes and the oil price increases, the CEGB's position had crystallised. The Board was convinced of the competitive nature of PWR generation costs; their advanced state of development and their inherently simpler construction procedures with a high degree of prefabrication (Various interviews and Williams, 1978, p.207).

The manner in which the Board presented its change of heart from no new orders to a large programme caused some considerable bemusement for MPs and others, especially when Hawkins told the SCST in 1973 that the Board had changed its figures before the Israeli-Arab conflict. As Williams notes and as is clear from the evidence (1978, pp.208-210, HC73-III respectively), Hawkins received a rough ride from the committee. The other members of the CEGB and the senior engineers were also convinced of the inherent difficulty associated with the AGR and believed that to adopt the SGHWR would be folly. It seems clear that Hawkins was reflecting the advice of his senior colleagues when he argued the need for independence from coal and oil via American LWRs: there was however some disquiet within his own ranks at the manner of his announcement, or more specifically the size of the projected programme.

On the whole most Board members had come to the conclusion that the troubles with the AGR meant that to maintain a strategic flexibility in supply in order to guarantee continuous generation, the CEGB had to build PWRs. Part of the reason lay in the severe labour troubles the Board was experiencing at all its sites. Most of the AGR was built on site, whereas most of the PWR could be prefabricated in factories where labour was more easily controlled, then the finished components could be shipped to the site of the station and welded or bolted together. The difference between the AGR and the PWR in terms of final output and cost of fuel tended to be marginal, but the construction factors put the PWR ahead. The SGHWR, however:

"...was one of those things that were out of time and out of place. Now if in the early years the SGHWR had been pursued vigorously and had more political sponsors, it could have appeared earlier and then it would have been a very serious contender. But it was like somebody suddenly bringing out the original Volkswagen Beetle now. You'd say 'Well, very nice, good attempt: but we have better models now, you know," argued one Board member.

Thus the CEGB's most senior engineers were committed to the PWR and were resolutely opposed to the SGHWR, which they considered already obsolete.

The manufacturers were also pressing for the PWR to be adopted, particularly Arnold Weinstock of GEC. GEC engineers had as their function to seek to produce cost-effective options within the designs approved by the NII. The case for the adoption of the PWR was an example of this and it is noticeable that the GEC managers and engineers came to a similar conclusion to the CEGB for similar reasons, but from different organisational motivations. The technical considerations when constrained by commercial necessity tended to produce a uniform case for the American system; engineers operating within or from a commercial perspective consistently opted for PWRs. It was this commercial awareness that caused Marshall to champion the adoption of PWRs; he believed that research should be slanted towards the products that were marketable by a nurturing of the entrepreneurial spirit which he believed to be latent within the technocratic ethos (also, Atom, 210, April, 1974). For GEC this meant that the PWR was the more desirable system for four basic reasons:

1   due to the higher content of factory-made components there is less site work and this therefore facilitates management control and quality and cost control;

2.   the main components are replicable;

3.   it appeared to be cheaper than British options;

4.   it was (and remains) the only type of reactor that British manufacturers had (and have) any real hope of exporting.

These views remained those of the manufacturers for over a decade and were forcefully put to the Energy Committee in 1980 (op.cit.). Thus, despite the suspicion in which the manufacturers were held by the CEGB following the AGR programme, there grew a common belief in the need for the ESI to adopt the PWR, although for different organisational goals.

To the problem of why an engineer and manager as distinguished as Tombs should break ranks with the rest of the ESI and advocate the SGHWR, there are perhaps two possible interpretations. First, that he genuinely believed in the British option and a mixture of nationalism, hard-headed realism (in regard to need) and engineering expertise made him resist the American reactors and champion the British one. A less charitable opinion is held within the CEGB and by some MPs. This view sees Tombs as deliberately setting out to seek personal aggrandisement, seeing the SGHWR as a method of doing this. One man who was on the Board of the CEGB at the time argued:

> "Now what old Frank was doing was backing the SGHWR and the AGR because that allowed him to be identified as a separate entity...he wanted to be identified in the public arena as a man of independent opinion. His motivation was highly political, he looked around for a viable answer and then started plugging it. Up there in Scotland he was a little nobody stuck on a little patch and he wanted to come South. The final decision was not a technical one, it was a political balancing of forces."

This was a view shared by many at the time and with hindsight. Certainly his stand earned Tombs promotion to the Chairmanship of the Electricity Council at a time when the Labour Government were determined to reorganise the industry and make the Council the most powerful body in practice, a move scuppered by the fall of the Government in the election of May 1979. The truth probably remains somewhere between the two views as even long after that debate and his elevation, Tombs remained a passionate advocate of the AGR and SGHWR and an equally devoted critic of LWRs.

Varley was subjected therefore to the barrage of conflicting advice outlined above. His own advisers could not agree and there were obviously some divisions within the civil service, with a majority of officials becoming convinced of the CEGB's and the manufacturers' case. Eventually he decided that since in the opinion of the trade unions a SGHWR would provide jobs and since there were safety questions still unresolved, he would opt for the SGHWR. One CEGB manager had a vision of him going home for the week-end with all the papers and then coming back and saying 'I have the answer'. He believed Varley chose it because it was British rather than be accused of buying in 'American technology' and therefore undermining British R & D. He went for the safe political bet, although it was still a technical gamble.

The evidence tends to suggest that this widespread and somewhat cynical technocratic analysis is the correct one, although it took Varley a good deal longer than a week-end to reach his very difficult decision, a decision made especially hard by the deep divisions amongst those technically qualified to advise. At the time it was and with hindsight it is, difficult to conceive of how those divisions could have been ameliorated, organisation rationality being such that it determined professional advice.

The Aftermath

The SGHWR was a dismal failure: in August 1976 Benn announced that due to technical slippage and mounting costs the SGHWR had been deferred, and in January 1978 work on the SGHWR was formally abandoned (Williams, 1980, pp.241-251). Some £50M had been spent on its development and yet it was not anywhere near to being completed; as an interim procedure Benn allowed the CEGB to place an order for another AGR at Heysham and the SSEB to build another at Torness. The reason there was a return to the AGR was that the performance of the stations finally built and working had been creditable and it was apparent that many of the design faults had been rectified. Furthermore, the manufacturers desperately needed new orders to remain viable and the safety problems with the PWR were still not resolved. Benn however alluded to some considerable pressure being put upon him in 1978 to adopt a PWR (HC 114-ii, p.394).

At least one AEA Board member considered the whole SGHWR saga to be a strange affair. He argued that the unforeseen technical difficulties had led Hill to believe that the cost overrun would be quite considerable so that he went to the Minister and said that there had been some difficulties and that it was going to cost a lot more than originally thought; therefore the best thing to do was to scrap it. He came to the decision that he had a 'moral duty' to tell the minister of the problems and to recommend cancelling the project.

This was a meeting which infuriated Benn. As an advocate of British technology and an unswerving opponent of the PWR he realised that this meant either a return to the AGR (and a struggle with the manufacturers and the CEGB) or an adoption of the PWR which he did not want. As was shown above in regard to the 1967 decision, Benn had placed a great deal of faith in the technical advice of the AEA throughout his time as minister responsible for nuclear energy; a recommendation to drop the SGHWR went against the best advice received two years previously from the Authority's own experts. As Williams points out (1908, p.244), the Minister candidly told the AEA that their recommendation undermined the credibility of British R & D and put into question the AEA's future credibility, particularly with regard to the FBR. In effect, the decision to recommend the cancelling of the SGHWR led to the vindication of the CEGB and the advice it tendered to the Whitehall machinery became that which was accepted above the options preferred by the AEA.

Despite their protestations, the Authority were in disgrace for their role in the SGHWR decision (AEA Annual Reports, 1977-78, pp.8-16; 1978-79, pp.5-19; 1979-80, pp.5-12). Benn did not want to order a PWR but the result of the cancellation meant that he had to give the option to the CEGB after the necessary safety studies had been completed and after the ESI had built the two new AGRs ordered in 1978. The civil service was united more than before in its support of the CEGB and manufacturer's case; previously, although a majority had been in favour of the PWR, there remained some firm advocates of the SGHWR. The decision to opt for the SGHWR had not only caused dismay in the CEGB, but it had even prompted one of the Under Secretaries in the DEn to resign: he argued that he could see this becoming another Concorde and that was something he did not wish to be associated with.

The aftermath of the cancellation united the civil service into such a firm advocacy of the PWR that Benn was to tell the Energy Committee in 1980:

"When I rejected the advice of my own officials they declined, for a period, to assist me in writing a paper containing the views that I held." (HC 114-ii, p.421).

This no doubt prompted his later remarks concerning the lack of accountability in the nuclear policy-system, arguing that 'it is not always possible even for Cabinet Ministers to discover the truth' (1984, p.5) and that:

"...the longer I have had contact with nuclear power the more sceptical I have become about the claim made on its behalf by those who present it as the answer to our major energy problems of the future." (HC 114-ii, p.398).

Benn in particular, therefore , felt that technocratic control of decision-making was a major influence in poor decisions being taken as those decisions were taken on dubious assertions using inadequate information. By 1982 he had become convinced that the arguments used by the CEGB and others to acquire large programmes of PWRs were a cover to defeat the power of the National Union of Mineworkers; to bolster the profits of major companies and to a certain extent to supply material for atomic weapons (1984, pp.4-9). His evidence to the Sizewell Inquiry was particularly revealing for an understanding of the way that decisions in the field are arrived at. He argued that:

> "The thing has got, in my opinion, much too sophisticated in the presentation down to three points of decimals, whereas actually ministers are serious people and so are civil servants and generals and Generating Board people; and they sit down and say "This is what we would really like. Let us see how we can present a good argument that makes it look as if that is the only course we can pursue". Now that is real life." (1984, p.18)

A view he later further elaborated by arguing that:

> "everybody who works in the field of high technology or works in the field of industrial development will have very highly reputable statisticians, mathematicians, economists and scientists working for them. But I must tell you that the main thrust and drive of all policy decisions rests upon the objectives and assumptions and beliefs of those who have responsibility for taking them, and to imply that there is a sort of vacuum of intention at the top until the computers have clicked out to the tenth point of decimals what would be the most economic conclusion...would be to misunderstand the political process." (1984, p.79).

It is clear, therefore, that with a knowledge born of painful experience, Benn was acutely aware of the conflicting goals of the organisations and individuals pressing him in the aftermath of the SGHWR decision and the problems of its development.

If the lessons for the decision-makers were in some ways a repeat of those learned after the AGR decision, then the lessons for the study of the role and influence of technical professionals in decision-making are also in some ways a reiteration. For the policy-makers the cost of nuclear nationalism at the expense of rational commercial/technical assessments was underlined. Like the AGR, the SGHWR simply was not the reactor capable of producing what they required of it. As a result the CEGB quietly let it die by calling for ever more stringent parameters and refusing to invest any more money in it. They did not exactly sabotage it, but when it began to drown under the weight of their requirements they refused to throw it a lifeline. The affair permanently reduced the status and influence of the AEA within Whitehall and conversely replaced it with the CEGB as the Government's chief nuclear adviser. For the politicians the affair reiterated the need to view the advice proffered by technical experts in the light of the organisational commitment of that expert (one of the rare exceptions being Lord Marshall, whose advocacy of the PWR Benn described as being as permanent a part of the London scene as Nelson's column (1984, p.36)).

The organisational bias of technical experts and their close identification with the goals of the employing body is a factor again underlined that is of importance for the study of the influence of professionalism in policy-making. The dual allegiance analysed in the work above and found in the case studies below was found in evidence, but the technocrats used their technical, their professional, expertise as a support for their organisational allegiance. As a result, when the different organisations were pressing for contrasting decisions it appeared that there was equal and ample evidence of a highly technical nature to support both sets of experts, as indeed there was. Benn argued, they started at the level of the general, that is the organisational intention, and then went off and found the evidence to support it. That is perhaps a too simplistic view, the organisational goals were formed by the detailed studies of the teams of experts, particularly in the CEGB and, as the next section illustrates, the Board's policy was the policy of its experts; it did not simply tell them to go off and find the evidence to support a Board viewpoint. What did happen was that the various organisations operated within the framework of broad general directions, such as the CEGB's brief to provide adequate supplies of electricity at the lowest possible price. Its interpretation of this brief then led to a commercial outlook that fostered a similar a priori viewpoint in its experts, along with an acute awareness of the power of OPEC and the NUM. These a priori interpretations of their role led the engineers to suggest options not always in accord with the wishes of the AEA or the Government and vice versa. Again, therefore, there is evidence of a dialectical process at work in that the influence of the technocrats upon policy was through their influence upon their employing organisation, which in turn caused them to operate within a particular worldview. Senior technocrats, like Lord Marshall and Sir John Hill, were influential by their interpretation of the political goals of the Government and the relationship of the technical options to those goals. In turn this interpretation coloured the advice they gave to the politicians.

When viewed together with the AGR decision, the SGHWR choice illustrated several further aspects of professionalised policy-making and professional power that were analysed above in Chapters 2 and 3. Specifically it was shown that when they are situated within and dominate the bureaucracies of organisations that are integral to the policy-system, the technocrats behave in the manner of other professionalised occupations. That is to say they use their power to define problems in terms that suggest technical solutions best realised through the exercise of their own professional expertise, as in the instance of the AEA's advice, or else which optimise the realisation of organisational goals and because of the extent of technical domination within the CEGB; and because the CEGB is the cultural centre for those occupations associated with its purpose, organisational goals are also occupational goals. The power to define needs within their own technical parameters facilitates the ability of the technocrats to define the available alternatives and therefore the allocation of resources, powers which the analysis of Chapter 2 (as interpreting the work of Wilding) illustrates is a necessary prerequisite to the assertion of professional power per se; the power, that is, to direct policy and people and to achieve occupational work autonomy. It also tended to support the hypothesis that professionalism increases the autonomy of single issue agencies in that the technical expertise resident within these agencies represents society's stock of that expertise and the ability of representative institutions to challenge the options preferred by the agency is severely limited. Furthermore, the professionalisation of the policy-system tended to centralise the decision-making process and to produce an implementation-skewed process of policy definition, factors particularly apparent with regard to the motivations of the CEGB and manufacturers.

The AGR and SGHWR decisions also showed, however, the limitations of professional power, particularly when the professions are technocratic, quasi-professional or professionalising occupations. Unlike the analysis of Wilding regarding the medical profession, the techno-crats of the nuclear policy system were continually divided in their goals and advice. This confirmed the analysis of Chapter 3 which saw the technocrats as divided professions, as groups of occupations in varying degrees of autonomy and professional awareness, 'engineering and 'science' being generic terms applied to a plethora of disparate occupational groupings. The situation was further complicated by the dual nature of technocratic allegiance and the necessity of positioning within bureaucracies, the collective nature of the professional product being apparent in these occupations (unlike medicine and law where the client relationship obscures the reality of the collective structures) and the complexity of organisational motivation qua the individual experts interposing upon their wider professional goals. Indeed the highest expression of their professional identity is to do their best for their employer. This naturally leads to the spectacle of professional fractures and competing groups of professional experts tendering contrasting opinions and advice to the policy-arbiters of the political system. Such a situation is rare in the social professions where professional allegiance is placed above that of loyalty to and identification with the employing organisation: indeed, where the majority of the profession are self-employed such a choice does not even exist. The result is professional unanimity and a unity of advice and purpose unknown throughout the technical occupations.

Unanimity facilitates the power of the social professions because there is little alternative advice to which the policy-arbiters can turn, therefore technical options proffered by the professions assume the aura of being the only viable technical alternatives when in fact they are frequently the result of selfish professional calculations. Such a collective unanimity does not exist within the advice proffered by the technical professions, the above examples showing that this allows a greater degree of political influence than would be possible were there no conflicts of technocratic opinion. The organisational nature of technocratic advice therefore facilitates the politicisation of decision-making, the politicians choosing the technical option most in sympathy with their political goals. The influence of the technocrats lies in their power to formulate the options, their technical expertise being the medium through which political goals are realised and through which the alternatives are interpreted to the policy-arbiters.

Technical professionalism in exposing the schisms within the occupational communities and the alternative options available also expose the contrasting motivations within the policy-system and thereby illustrate the sham of apolitical decision-making. That is, by showing the political considerations behind the technical decisions, technical professions are able to show that 'rational' techniques and 'non-political technical decisions' are in fact highly political in content and in potential. The divisions within the experts allowed politics back into the nuclear field and the energy field from whence the experts had tried to banish it, as such the potential for increased accountability was to be glimpsed. Such an increase would necessarily reduce the autonomy so beloved of professional groups and this accounts for their belief in the Morrisonian approach called for by Hawkins and sought by the others that would allow decisions to be taken on 'technical grounds' that sought the best 'technical' solution. The decision to pursue the PWR option after 1979 further illustrated the political nature of decision-making and the dependence of the technocrats on presenting their technical options in terms favourable to the politicians. The 'commercial' advantages claimed for the PWR being heavily infused with broad political calculations and ideological propensities.

THE PRESSURISED WATER REACTOR

On 18th December 1979 David Howell, the Secretary of State for Energy in the new Conservative administration of Mrs. Thatcher, announced the Government's energy programme. It represented an investment of about £15Bn over ten years (current prices) into PWR technology, or ten power stations of about 1200MW each over a decade, beginning with Sizewell B at Suffolk. The announcement appeared to be a vindication of the long campaign waged by the CEGB and manufacturers for the LWR option, and it was greeted with dismay by the coal lobby and environmentalists (Guardian, 19/12/79; Sweet and Coote, New Statesman 20/11/81). Just as the AGR and SGHWR decision had been political, so the PWR was chosen for a variety of factors including the Government's direct response to he perceived militancy of the NUM, as leaked Cabinet documents made clear (Sweet and Coote, ibid.).

Despite their apparent victory the CEGB were unhappy with the announcement, correctly interpreting it to herald a period of heightened public interest in nuclear power, much of it inflamed by the fears of the environmentalists. Public interest meant political interest and that led inevitably to many decisions, that would normally be left to the utilities to make on a day-to-day management basis, being instead the subject of ministerial scrutiny and intervention: the antithesis of professional autonomy. One Board member said that Howell's announcement, against the strongest of CEGB advice, brought 'all hell down on us'. Specifically it aroused the combined wrath of groups such as Friends of the Earth and the Town and Country Planning Association who were concerned about the environmental impact; the Campaign for Nuclear Disarmament and Tony Benn who were concerned that nuclear products from British reactors would go towards American nuclear weapons under the 1958 Mutual Defence Treaty and the agreements reached between President Kennedy and Harold Macmillan (as later ratified by Harold Wilson) (Guardian, 8/3/85; New Statesman, 17/6/83); and the NUM and its allies in the coal lobby (Benn, 1984). CEGB members were angry at the announcement of a programme, arguing that such a programme was neither sought nor necessary and that Howell had inserted the clause outlining the intention of a programme for purely political reasons of his own. The Board was not allowed to veto the statement and it was strongly criticised by opposition groups, causing considerable discomforture to the CEGB.

The statement itself appeared in a CEGB publication of 1981 labelled the 'Thermal Reactor Strategy'. In practice, the statement by Howell was an endorsement of the policy enunciated (and since repudiated) by Benn in January 1978 when he said the Government's aim was to 'establish a flexible thermal reactor strategy' for the UK and ESI 'in the light of developing circumstances'. As well as discontinuing work on the SGHWR the Government allowed the SSEB and CEGB to begin work on one more AGR each as soon as possible and 'accepted the need to develop the PWR option in the early 1980s', and that providing all the necessary design work and consents were completed, the ESI 'would order a PWR station' (CEGB, 1981, p.7). This policy was endorsed by Howell, the Government arguing the electricity supply industry had advised that even on cautious assumptions it would need to order at least one new nuclear power station a year in the decade from 1982, or a programme of the order of 15,000MW over ten years.

But it was this statement that the CEGB fundamentally disagreed with in arguing that it might be necessary to obtain one site a year in advance of need on the grounds that it took a long period of time to acquire planning permission. This strategy, it was hoped, would prevent any possible bottlenecks in the future, but a programme was not envisaged: on the contrary, the CEGB wished to consider each site on its merits within a flexible and continuing policy of review of possible future need. The DEn argued that the CEGB and the opposition groups misinterpreted Howell's speech and pointed out that the Government was not calling definitively for a large programme, only that they wanted provision for a programme if necessary. A senior MP disputes that interpretation of Howell's speech, saying it was not so much a projected programme as a 'statement of intent'. He pointed out that in his view the country needed the increase in nuclear power for strategic reasons, the Cabinet having decided that the country could not afford to be dependent upon coal and the NUM. The Strategy said:

> "It should be emphasised...that the reference to the Government's 1979 statement to the 15,000MW of nuclear plant over 10 years should not be construed as a commitment to any one type of system or level of ordering. Such a commitment would be wholly inconsistent with the concept of flexibility which is the key feature of the strategy". (op.cit., p.8).

Thus the opposition (and indeed the CEGB) somewhat misinterpreted the extent of the Government's commitment to the PWR. Indeed the moderate strength of that commitment was amply demonstrated in March 1987, when the Secretary of State for Energy, Peter Walker, followed his announcement that the Government had approved the construction of a PWR at Sizewell (Guardian, 2/3/87) with permission for the Board to build two new coal-fired stations. Some of the hyperbole surrounding the 1979 announcement was a simple ploy to prepare the ground for a modest programme of PWRs, Sizewell being followed by a PWR at Hinkley

Point and possibbly two or three more. The fact that there was a clear and emphatic expression of support and intent for the PWR is what was so important about the failure of the SGHWR and the shift towards the acceptance of 'American' technology by the political system.

The Failure of Nationalism

With the decision to cease development of the SGHWR, the case for the British reactors appeared lost, despite the improved performance of the AGRs and the decision to build two more. The latter move was as a result of strong pressure by the industry for orders, as Benn told the SCE in 1980; he ordered not only two new AGRs immediately but also the huge Drax B coal-fired station in Yorkshire to protect the turbine and boiler-makers and to keep the coal industry healthy (HC 114-ii, pp.426-434). A member of the Committee (Stoddart) asked incredulously whether this meant that:

> "...we would be building new power stations, although we already have available capacity, just for the hell of it and just to keep the industry going" (ibid).

Benn retorted that the costs of allowing the affected industries to atrophy would be almost incalculable, certainly within the realm of £20Bn. worth of imports in the short term alone when the time came to replace old plant. This illustrates the guiding motivations for the Government's decision-making process. It tends to support the analysis above that nuclear power was viewed by successive governments as an adjunct (albeit a tremendously important one) to broader general goals, whether military, industrial or economic. The experts most successful at shaping policy were those most successful at presenting their advice as being in accord with those broad aims. Benn's candid expositions as to why certain options were chosen clarify the complexity of motivations inherent to the civil nuclear decision-making process.

The experience of Benn is particularly interesting because from 1976 until 1979 he found himself at odds with the officials in his department, the CEGB's senior management and several of his senior colleagues in Cabinet. The pressure put upon him to adopt the PWR was considerable. Price has argued that experts can only contribute to making a decision if politicians have first specified a consistent set of objectives and they are the intermediaries between abstract knowledge and direct action (1965, pp.124-134). The adoption of the PWR is an example of this in almost reverse order in that from the 1960s commercial and organisational rationality suggested the adoption of the PWR and this was frustrated by the politicians' desire for a British option to be adopted (as advised by the AEA). Benn himself, because of a mixture of concern to retain British jobs, nationalism and continuing doubts about LWR safety, gradually found himself isolated. Starting from the CEGB, the experiences of the AGR and SGHWR gradually convinced the civil service and other ministerial advisers (including the Chief Scientist whom Benn had appointed to replace Marshall, Herman Bondi) that the PWR option was the best solution to future problems of energy supply. This is not an example of technocrats and civil servants imposing their decisions upon the Minister against his will, rather that Benn himself had become somewhat isolated within the political system on this issue and was forced to compromise to some extent. The options presented by the technocrats may well have clashed with Benn's personal preferences, but they remained closely trimmed to the commercial requirements of the industry as a whole.

Benn complained about considerable pressure because as Minister he was out of step with the near unanimous opinion of his advisers and his political colleagues. It was this lack of political support which isolated him when he made his decision. In 1974, Varley was faced with a divided set of opinions and a Cabinet predisposed to the British option, the experiences post that decision were to unify support for the PWR as the option likely to bring the greatest economic benefit to the country. Benn was viewed as an obstacle to this by the civil service and the managers in the CEGB and the manufacturing firms. It was as if they had moved on and left him behind, although his personal views were becoming increasingly ill-disposed towards nuclear power, recognising not only environmental dangers but also the threat it posed to the coal industry. He said that he was:

"opposed to the PWR but could not sustain the argument (in Cabinet) that you should rule it out altogether" (1984, p.26).

He announced his 1978 statement as being the collective decision of the Government and it clearly represented a compromise between Benn and the opposing interests in Cabinet who wished to adopt a more hard-nosed commercial stance.

Although Benn was forced to accept the PWR as an option in the 1978 Cabinet meetings from which the statement emerged, he did manage to prevent the immediate adoption of the reactor by the CEGB, much to the latter's irritation. The statement reflected a fudging compromise, yet it did open the door to the emphatic acceptance of the PWR in 1979. Certainly the civil service by 1978 were in no doubt that they should advise the minister to pursue the PWR route. The divided loyalties of 1974 had coalesced into the united front described above. Benn recalled that:

"It was the first time in my ministerial life that I had a meeting with all my officials, and they were unanimous under the Permanent Secretary that we should adopt the PWR. I declined to accept that, and as a result they declined to draft a paper for me to say what I wanted. It was a very extraordinary example, and so I got my advisers with the help of one official who was prepared to put his career at risk, and my Department then made its paper available to the CPRS who put it in a paper of their own... It was a most interesting example of a Department in effect going on strike against its Secretary of State. They simply would not prepare a paper because they did not agree with the decision that I had reached" (1984, p.19).

Meacher regards this as an example of 'Mandarin Bureaucracy' and a threat to the democratic process as represented through the duties of Ministerial Responsibility (Guardian 14/6/79). Plowden, however, views it as an essential part of the workings of democracy, astute and aware civil servants with a public conscience acting as a necessary check and balance upon Ministers (letter to The Guardian, 20/6/79). This was to be an argument rehearsed again in the aftermath of the Ponting affair.

The PWR decision illustrated the almost federal nature of decision-making within Britain's unitary system of Government, the various interests battling amongst each other from 1962 until uniting in a nearly unanimous agreement on the need for the PWR: the AEA recommending it in order to free resources for a concentration upon the FBR after 1980; the CEGB wanting it for the reasons outlined above and below, namely that they believed it to have manufacturing advantages over the British reactors with which they had experienced so much trouble; the manufacturers wanting the LWR option because it allowed them the only opportunity they were likely to get to export in the nuclear field; the civil service were convinced by a combination of the above arguments, mainly though the realisation dawned that by forswearing the LWRs Britain had isolated itself from the rest of the World's nuclear technology. This was a path fraught with expensive possibilities in an era where the need for international collaboration impressed itself forcefully upon the consciousness of the Whitehall structures. The experience of Benn showed the relationship of Ministers to the technocrats from a political viewpoint, rather than the technocratic viewpoint which was analysed above. Benn claimed to have difficulty in getting political decisions implemented by the CEGB: indeed he even had difficulty in getting his questions answered:

> "When I asked Sir Arthur Hawkins what was the price he paid for coal, he said quite bluntly and plainly, because he was a blunt and plain man, "That is a management decision. You have no right to ask me", and he actually laid upon the table between us the statute which he claimed authorised him not to answer the question." (1984, p.17)

In fact, by not reappointing Hawkins, Benn was able to impose his Ministerial will, but it is useful to note the conflicts and struggles that existed between Ministers and the organisations they were expected to control. The process by which technical opinion informed and shaped organisational views, which in turn influenced political advice, was a dialectical process. The CEGB formed its views due to the brief it had been given by the politicians; it was a commercially- minded organisation and its engineers took their lead from this. A minister with less self-assurance than Benn, or someone who lacked his convictions, could easily have been swayed by the arguments of the technocrats and ordered a large swathe of PWRs. The fact that neither he nor Varley was so tempted illustrates the point that the final decision lay as always with the minister, but that the weight of commercial, political and technical evidence was making such a position as that followed by Benn, in a Government committed to nuclear power, untenable. Increasingly, to accept nuclear power meant to accept the inevitability of the PWR, unless there was to be new evidence of an unanswerable kind presented. Within the confines of the ESI in Britain in the late 1970s and early 1980s, the only option to PWRs appeared to be no new nuclear power stations, a position only altered between 1983 and 1987 and the impressive showings of the new AGRs as they finally began to work, a performance tarnished by yet more revelations of design and operation failures regarding the control rods at Heysham and Torness in 1987 (The Observer, 1/2/1987).

In 1978 Benn looked for technical evidence that would allow him to pursue the political options that he was committed to, that is more coal stations and British-built nuclear reactors. This was exactly how Varley had behaved in 1974, but whereas Varley was able to cover himself with the figleaf of Tombs's and the AEA's advice, Benn was bereft of even this expert cover for political decisions. Indeed, within the political system itself he was isolated and the compromise decision of 1978 which opened the door to the PWR was hammered out. The unusual unanimity of expert opinion when combined with the civil service's political calculations concerned with other aspects of economic policy, made the choice of the PWR irresistible, until Three Mile Island and Chernobyl that is when other political calculations intruded. But it must be emphasised that it was organisational needs, not technical rationality that governed expert advice (except that of Cottrell who, now retired, continued to warn of the safety problems connected with the PWR) (HC 114-i, p.120-125) and essentially the organisations that were situated within the nuclear industry were concerned with making money, either directly as in the case of the manufacturers, or for British industry per se, as in the case of the ESI.

The Rationale for the Choice

For the CEGB the rationale for the choice of the PWR came from its experience with British reactors and more importantly, from its 'Thermal Reactor Strategy'. Benn argued that the CEGB edged out the SGHWR and that anything which 'got in the way' of ordering PWRs was 'somehow edged out again' (1984, p.45). Upon his appointment as Chairman by Benn in 1977, England ordered a review of the thermal reactor options available to the cEGB, with the expressed intention of 'discovering' the optimum system and one which could be presented to the politicians as the best system. It was clear to England that the CEGB's commitment to nuclear power had come because of the militancy of the NUM which combined with the oil crisis had again demonstrated the vulnerability of Britain's energy supplies. Also, the performance of the Magnox reactors had been very impressive; operating on base load they were to become what Hawkins called 'the workhorses of the CEGB' (CEGB Annual Reports, 1972-3, 1973-4, 1974-5). England also argued that because of inflation nuclear power by the mid-1970s was for the first time genuinely economic compared with coal and that the comparative costs of fuel made nuclear stations competitive , a view that Sweet fundamentally disagreed with (1982). England appointed another Board member, Dennis Lomer, to head the investigation of the different reactors available. Yet although the assessment was a penetrating investigation, the Board was already allowing the SGHWR to die a quiet death in the manner described by Benn. In 1977, England realised that there was no need for the CEGB to order a new station immediately on grounds of supply, although the manufacturers wanted work; and he felt that 'a long reassessment of policy' was called for. A thorough analytical study of the CEGB's requirements followed.

It had become clear to the Board that by burning 80 million tonnes of coal a year they were dependent upon the miners and that in order to achieve fuel diversification nuclear expansion was necessary, North Sea oil and gas being both expensive and not expected to last much into the 21st century. When England told them to recommend a reactor a member of the assessment team recalled they established:

"...a meeting with the Directors General and the Chief Officers such as the D.G. of Research, and Generation and Design, Health and Safety etc. It was decided that we would set up a steering committee and we then gave them the terms of the reference that they were to distil out the facts and produce a report, but under no circumstances were they to include in their report any recommendations or any conclusions. Lomer used to have periodic meetings with the steering committee, and the committee itself set up small working parties, for example on the advantages and disadvantages of CANDU, the PWR, AGR and Magnox."

When the report was finished and Lomer was satisfied with it, he took it to the Board and the Board briefly discussed it before setting aside a whole day where they got the steering committee to meet with them and discuss the whole thing. At the end of the day there emerged a policy to put top priority onto finishing off the AGRs that were under construction and immediately set in hand a new design study, update the Hinkley design which was just about to be commissioned and if that was satisfactory then order another AGR and at the same time do a detailed study with Westinghouse and the HSE or the NII on the PWR. They wanted to satisfy themselves and the NII that they could safely build a PWR with a thirty-year life-span.

Both Lomer and England eventually managed to convince Benn of this strategy and agreed with him that a large programme was not necessary or desirable. The assessment was the first time that every level and Group within the CEGB were brought into a top-level policy-making group. The assessment was not rigged, but it did confirm the CEGB's predisposition for LWRs, showing them to be cheaper and easier to build; it was part of a proven world-wide system with 'all the bugs ironed out'; it would provide cheaper electricity; it would diversify supplies; the Board could order it when it wanted it without any development costs; it was easier to construct and due to the high prefabrication element it facilitated site labour control and therefore helped prevent the sort of delays experienced over the AGRs and Magnox. A viewpoint substantially confirmed by Layfield at the Sizewell Inquiry (1987, Chapters 2,3,47,90,108 and 109).

If the Thermal Reactor Strategy confirmed the CEGB's propensity towards water reactors it certainly satisfied the manufacturers likewise. GEC and the others had retained the views expressed at the time of the SGHWR decision (see above) that the PWR was the only system that British manufacturers had any real hope of exporting due to the fact that the rest of the world had adopted it. Also, its construction costs were lower due to the replicable nature of the main components and the prefabricated nature of those components facilitated management control and quality and cost control. Of the manufacturers only NEI favoured a continuation of the AGR, arguing that it could be developed to compete satisfactorily with the PWR (HC 114-ii, pp.307-315), a view shared by the continuing individualistic stance of the SSEB (ibid, p.327). Most of the others involved in the industry welcomed the switch to PWR technology. Weinstock told the SCE that:

> "...my views on AGRs come from several years' experience of having to build them and that experience is enough to put anybody off them" (ibid, p.479).'

It was these commercial considerations that had convinced Marshall of the desirability of switching to water technology. He became convinced whilst still at Harwell that it was a mistake for Britain to pursue an individualistic policy with regard to reactors, believing it would cut UK technology and UK manufacturers off from the world market and international collaboration. Indeed, previous to Benn's announcement opening the option for a PWR and giving permission for two new AGRs, Marshall had tried very hard to convince him of the need to build a PWR, citing the willingness of the Shah of Iran to fund such a switch as a further good commercial reason for doing so, pressure that further alienated Marshall from Benn and hastened his replacement by Bondi. Benn recalled:

> "(Marshall) came hot from Teheran and told me that if only we would order the PWR the Shah would buy half our industry, and when I said "How come you were discussing it with the Shah?" he said to me "Because I am the Shah's atomic adviser". I said "I did not know that. I thought you were mine", and he said, "Well, the AEA have some arrangement with the Iranian Government". I understand Dr. Marshall's position completely". (1984 p.36).

By 1980 Marshall had refined his position to favour the PWR over all other types of reactor, although he believed that both AGRs and a PWR should be built in Britain in order to obtain a full comparison (HC 114-i, pp.94-99). His main argument for the PWR remained that the rest of the world had them and therefore any problems which emerged would have world-wide resources devoted to them; but he said that he would accept AGRs if the Government changed its mind, saying 'I try to be an unemotional technocrat' (ibid). By this time Marshall was Deputy Chairman of the AEA and the Authority was fully committed to the adoption of LWRs, even arguing it should, with hindsight, have been done much earlier.

Even the replacement of Marshall by Bondi did not alter the advice being fed to the Secretary of State, Bondi echoing Marshall's views. A scientific advisor of the time recalled:

> "There was very great pressure put on Benn to go PWR. Outside and inside the industry people believed that the PWR would have considerable export possibilities. Secondly, there is the

perfectly sound argument that a PWR needs less outdoor work than an AGR and outdoor labour in this country has always been the most difficult one to control and to keep productivity anywhere in line with estimates. I would not totally dissociate myself (although I'm not a PWR man basically), in the nature of things an American/German/French and Japanese system has had a lot more brain juice poured over it than a purely British system. That's not a bad argument for the PWR".

Bondi's role provided a further insight into the influence of technocrats in Whitehall and their perception of the political process. The ethos of Whitehall is such that it is how a person is received, not what they say (within reason) that is important. That is, if senior civil servants perceive a person to be a political ally and their judgement has been sound in the past then they are likely to have greater influence than otherwise, regardless of their standard of knowledge; the relationship between Churchill, Cherwell and Tizard is perhaps an earlier example of this (Gowing, 1964, Part 1). One former Chief Scientist said:

"My role was to form a judgement and then explain it to a non-scientist. My children think I am a great but rather successful fraud; they are probably right, but fraud has its purpose. I would explain things to both ministers and administrative civil servants: <u>there is no great distinction there</u>. You may be talking in some fields therefore where you are an expert, some where you are a scientific generalist and some where you are an ordinary citizen: you've no greater knowledge than the ordinary citizen who reads the newspapers. But you happen to have the ear of the Minister or a senior civil servant. Here you must be very careful: the ethics of the profession rely on not mixing those roles and making quite clear when you speak without any qualifications that you express your views but you claim no particular standing for them".

In the role of Chief Scientists in Whitehall and their own interpretation of that role, there are obvious links to the Kuhnian thesis of the paradigmatic nature of scientific truth. The Chief Scientist quoted above, however, provided a severe qualification to this by his protestation that:

"I'm not a particularly moral person, but I do know that I'm not half clever enough to lie consistently and so I acquire honesty by default rather than for any other reason. It's a terribly difficult thing to lie consistently and I'm just not up to it".

It is not through 'lying' that information becomes slanted or distorted, but through the preconceptions and organisational loyalties of the experts; sometimes they are not even aware of this as norms are internalised; such is the way of professional socialisation. An example is where it was realised that even though Bondi was hired by Benn to provide alternative advice from that of the AEA and Marshall. He in fact actively sought that advice from Marshall himself (as did other senior Government scientists and engineers) considering it to be of the very highest standard and viewing Marshall highly as a fellow professional, skilled in a different discipline and whose advice, pertaining to his expertise, should therefore be sought by other experts and weighed favourably. A Kuhnian view is again glimpsed with regard to scientific paradigms (1974). It is possible to argue that when approaching a problem, therefore, the experts were unable to be fully objective. Indeed, their advice was bound to be value-laden with the goals of their employing organisation: and if it was not, awareness of the political situation, so necessary in the organisations involved in policy-making was lacking. Policy-options therefore tended to be the result of (and reflected) the organisational bias of the experts who formulated them. Benn learnt this lesson and the ESI's managers were particularly aware of it. Several senior technical experts also recognised this and claimed to take account of it when delivering or weighing advice.

Ultimately, the choice of reactor system has such a high political content that it cannot but be a political choice and this is accepted by all parties concerned, although profoundly regretted by some. One Whitehall scientist cogently summed up the view of many when he said:

> "Quite clearly in many fields there will be a beautiful technical solution where your Secretary of State will say 'You've been very convincing and the answer is no. I'll never get this through the House of Commons, it won't wash'. That's his business; that's what he's paid to do, to take that decision. Now clearly the more public interest there is in something then the more the decisions must take account of public opinion".

That is essentially what Benn attempted to say and what Varley had successfully argued previously. The weight of political considerations other than his own brief had become important by this time and therefore Benn was politically out-manoeuvred: it was alternative political arguments, not technical ones which swayed the decision. As Fey argues (see above Chapter 2), positivistic positions of claimed objective rationality are themselves subjective political positions that are usually conservative in content. The decisions for the LWR, taken as they were in such a charged set of circumstances, went some way to exposing the political nature of decisions within the technical realm. The concern with exports, profits, labour control and defeating the power of OPEC and most of all the NUM, all brought into sharp focus the value-loaded nature of the civil nuclear power decision. The PWR was the 'best' decision within the parameters of the ESI and the manufacturers; it was not such a welcome decision for the coal lobby, the civil engineering workers or the environmentalists. The political nature of the decision was further illustrated by the manner in which it was embraced and extended by the Conservative Government in 1979 and 1987.

Conservative motives and the PWR

The Conservative Government came to power in the Summer of 1979 elected on a platform of radical market policies. Essentially, they sought a transfer of power away from the corporatist structures of the Welfare State, especially those that had evolved in the 1960s and 1970s (see Smith, 1979). The theories, known collectively as the 'social market theory' (Gamble, 1981) and introduced to the Conservative Party by Sir Keith Joseph and the Centre for Policy Studies (1975), followed the work of neo-classical economists such as Milton Friedman (1962). Within the potpourri of Nineteenth Century Liberalism and neo-classical beliefs lay the foundations of a desire to curtail dramatically the power of the trade unions (especially the NUM), remove state subsidies for inefficient industries, and promote rapid economic growth through the medium of market-adjusted rewards and punishments.

The PWR was seen as a medium through which a disciplining of the NUM could take place. Leaked Cabinet documents (see above) plainly showed the Conservative Government's determination to use the expansion of nuclear power as the means through which they could loosen the miners' grip upon the economy. The administration was convinced by the arguments of commercial expediency espoused by the manufacturers and the CEGB, and England was soon replaced by Marshall as Chairman of the CEGB. The desire of the Government to put the nuclear industry on a firm commercial footing which favoured private enterprise accompanied the commitment to PWRs. Almost the whole of the Cabinet was opposed to the AGRs because of their construction record, and Howell had a hard fight to keep the two orders made by Benn before Labour lost office, the Government being convinced that orders made on purely nationalistic grounds without sound commercial reasons were an inherent reason for the weakness of British industry. Henceforth all orders were to be made on the soundest financial grounds possible; sentiment was banished and nuclear power had to pay its way as well as provide an alternative to coal. Eventually, the AGRs were confirmed in order to maintain the manufacturers until such time as they were able to tender competitive designs for the PWR. It was felt that the statement of intent was necessary to give confidence to private industry to tool up, it being no good to order just one; "that would be like ordering just one Rolls Royce," argued one MP. The Government had been convinced by the arguments of the manufacturers that they could export components profitably on the back of a domestic market and had only been excluded from the lucrative world market by the fact that Britain had isolated itself for so long. This was an argument which did not however ask why foreign utilities would suddenly switch to British suppliers when American, French and German manufacturers offered turnkey contracts, and in any case had infinitely greater experience in PWR technology.

A second factor accompanied the decision to endorse the PWR, a desire to infuse a greater commercial awareness and private participation into the public sector, something strongly resented by the CEGB. England's fate was sealed after the Minister for Consumer Affairs, Mrs. Oppenheim, instigated an investigation into the ESI by the Monopolies Commission (HC 315, 1981) and some fairly trenchant comments by the Commission regarding the Board was combined with his opposition to the Howell statement. It was only a matter of time before his retirement and replacement were announced, hence Marshall moved into his seat after he had served only one term in office. This appointment was in recognition of Marshall's technical and managerial experience, but also as a result of his strong advocacy of the PWR and of private enterprise and commercial success. The Government of Mrs. Thatcher was well aware of Marshall's commitment to the commercial exploitation of the PWR, and therefore they charged him with the task of establishing a framework for its construction which would favour the wishes of private industry (see also Kemp,1986).

The CEGB considered itself a successful commercial organisation without being taught its job by a 'newcomer' from the AEA and his allies in the Government and private industry. There was some considerable resentment at Marshall's appointment and exasperation at what the other Board members considered to be his lack of commercial awareness. One Board member at the time recalled that when the PWR eventually came to be built, Howell:

"...was persuaded by the private sector to say that the PWR would be built by the private sector and that NNC would have total project control. This meant turnkey virtually and also that the CEGB would have to stand aside and NNC would commit money and expenditure and the Board would pay".

Some members of the Board considered that this was a retrograde step; they felt that it was in fact a return to the old system that was responsible for the cost overruns and delays of the programmes and they resisted it furiously. Eventually they took early retirement. Marshall upon his appointment had spent the previous seven years on the Board of NNC and he tended to take his advice from them: the old feud between GEC and the CEGB broke out again, and the Board considered that under Marshall there was a return to a system that allowed the manufacturers to make profits at the expense of the electricity consumer without there being adequate control and supervision.

As the board member most responsible for achieving a measure of industrial and financial accountability from the firms manufacturing power stations, Lomer was the one who felt the strongest that tendering should be on a competitive basis for individual components (or that where necessary the Board could place an order directly with a 'good' firm) and that the direct placement of orders for those components should be backed by financial sanctions(Lomer,1985). Plans for the NNC-controlled manufacture of the PWR at Sizewell proceeded until well into the Public Inquiry into the site. The CEGB was then able to use the inter-vention and strong criticisms of the structure by the leading industrialist, Sir Alastair Frame, to invoke a restructuring that was very near the system evolved by Lomer (Kemp, 1985; Layfield Report,1987). Although Lomer did not entirely agree with the Frame/CEGB compromise, it was at least in his view better than that which prevailed before the Inquiry (Financial Times, 19/4/85 and letter 30/4/85, also Lomer, 1985, IGE communication 1254).

The highly political and strongly ideological approach of the Conservative Government, although at last giving a firm commitment to the LWR, also provoked some initial conflict with the organisational goals of the CEGB. This is not to reify the organisation, yet clearly the Board did possess an ethos different from that of other organisations. Although committed to commercial methods of proceeding, the Board was not concerned with serving a profit motive: it used the accounts as a method of measuring efficiency in providing value for money to the electricity consumers, not as a way of measuring a financial surplus. Indeed Chapter 8 illustrates that there was even some residual hostility to the idea of a public utility making profits as when the BNFL executives were negotiating prices for nuclear fuel with the CEGB they were asked 'Why do you want to make a profit?', as if it were a form of profiteering at the public's expense. The Board was a corporate body, literally and ideologically; they believed in a planned a controlled market, a guided and negotiated economy. Well-versed in the corporate structures and workings of the post-war economy, indeed a product of those structures and a vital component in them, the CEGB was hostile to the threat of the social market being extended to them, and according to their rationality it was a curious even foolhardy thing to attempt to do.

Ironically, in attempting to reduce state interference in the economy, the state/Government had to intervene directly in the workings of the CEGB as much, if not more than, the previous Government: hence the lament that each successive Government is worse than the last and they are all interventionist (see previous chapter). Such a political approach also underlined the reality of the true nature of technical decisions in society and exposed the facade of techno-rationality as the ideologically infused support for the dominant social and political norms that is its function. The 'best' technical decision varied according to the perspective of the organisation advocating it and the politicians at times were spoilt for choice. The coalescing of technical opinion around the case for the PWR was not an example of undue technical influence in decision-making; quite the contrary, it was itself the result of a rationality derived from the internalised norms of the technocrats. Those norms were the prevailing commercial, Liberal-Democratic ones of the society and organisations in which they were situated. The parameters in which they made their decision and formulated their options were set by the political system and the conflicts that occurred were about detail, not fundamental ideological issues. At no time was the 'enemy' the politicians for the ESI, certainly not in the same way as the environmentalist lobby was perceived. The technocrats were always on tap but not on top, and any mutinous stirrings were easily (and ruthlessly) quelled: their policy options were rigorously trimmed to reflect the propensities of those options and the interpretation of instructions for implementing policy decided upon.

Conclusion

The last four chapters have argued that the primary motivation for nuclear energy in the 1950s and 1960s was nationalism. First, it was the realisation that the scientific discoveries provided by the military projects could be harnessed for the purpose of civil exploitation. The engineers of Harwell and Risley sought to utilise the heat that was expelled as waste from the piles used to manufacture the radioactive products that were the components for atomic weapons. Out of this grew the concept of the Magnox reactors, low-temperature plutonium breeders primarily useful for providing plutonium, but also the precursor of the more advanced gas reactors. Forced by American isolationism after 1945 to develop its own reactor systems without the benefit of substantial enrichment processes, the British pursued the gas option as the only credible alternative. The Magnox themselves were given a boost when the shortfall in coal production and the Suez crisis showed the vulnerability of Britain's energy supplies. Economic returns on investment were not regarded as possible with the initial programme: it was a concern to safeguard energy supplies through alternative sources and to establish a viable British capability in nuclear technology that were the motivating factors.

With the second programme came the desire to exploit commercially the technology. The Oyster Creek break-through by the Americans and the threat they posed to the world export markets meant that Britain had to demonstrate the commercial viability of the AGR if it were to export the reactor and recoup some of the development costs from overseas orders. Indeed, it became apparent that due to the small size of the British market, exports were essential if the stations were ever to be commercial. The perceived need to show their export potential was a reason for the AGR appraisal; that and the need to find a politically acceptable solution to the near deadlock between the AEA and the CEGB, the latter wanting to explore the possibility of an American or Canadian reactor system.

The organisational rivalries were another factor in the decisions and the hegemony of the Authority in nuclear questions led to the pursuance of a nationalist path for far longer than a commercial approach, as favoured by the CEGB, would have allowed. Although they function within the parameters and set goals of their own organisations, the teams of experts also shape the goals of their employers, particularly in pure research bodies like the AEA, but also within a professionalised industry like the ESI. The attachment of the AEA to the gas technology was a reflection of the research choices of its scientists and engineers and their allegiance to each other, a factor which Hill later admitted caused some harm to the industry by preventing a switch to water reactors at the same time as the French, and thereby effectively cutting the UK off from the mainstream nuclear development in the world [42]. Price's belief that science can contribute to making a decision only if someone has first specified a consistent set of political objectives is not entirely borne out by the reactor decisions. Political objectives were rarely clearly expounded and specific but internalised, the only real exception being the Conservative Government of Mrs. Thatcher.

Yet as the writing of both radical and liberal analysts shows, in the post-war British state the whole purpose of political development has been a striving for and nurturing of a corporate consensus (Smith, 1979; Middlemas, 1979) and this is internalised within the decision-making procedures of those employed by organisations serving the state so that the decisions they arrive at are those calculated to advance the consensus and promote economic growth. Alternatives are not even considered as those framing policy options are concerned with their political and commercial acceptability: their rationality is a disciplined liberal rationality (Bachrach and Baratz, 1962; Habermas, 1979) that is confined within the advancement of the status quo. The result is that technical reasoning and scientific rationality are used to support political choices (Habermas, ibid; Wynne, 1982). To this end the ideology of 'professionalism' is both a form of controlling an occupation as the work of Johnson (1972) argues and a means of advancing the systematic upgrading and privilege/autonomy of specialised occupational groups in society (Larson, 1977; Wilding, 1982). As professionalising occupations, scientists and engineers are constrained by their 'professional' ethics to behave in a manner calculated to support the general mores of the state, and due to their dual allegiance this is reinforced by their loyalty (premier loyalty) to their employing organisation: an engineer is seen to be serving his profession when he is serving his employer to the best of his ability.

In this the technocrats were emulating the commercial and economic concerns of the politicians. Concern was to support British industry and the British economy. Even Benn identified the interests of industry and predominantly the private sector manufacturers as the interest of Britain as a whole (see above, analysis of his evidence to SCST and SCE). The profitability of private industry and its competitive ability in relation to the rest of the world was the reason that politicians wanted secure supplies of cheap electricity: any group of experts who could promise that were guaranteed the ear of the Government. In the nuclear sector per se, it was learned that British technology was isolated and the native market too small to sustain a viable and profitable industry; therefore commercial considerations demanded that a switch to the LWR take place. This became inevitable when the strongest supporters of British technology, the AEA which had developed it, decided under Hill and Marshall that linking with Europe was the only way in which to ensure that the UK continued to have access to the sort of market and funds necessary to sustain a first class effort. The election of Mrs. Thatcher did not lead to an alteration in the broad policy of the state with regard to civil nuclear power; it simply made more explicit the political underpinnings of that policy, especially in its dealings with the labour movement. The 'attack' on corporatism, such as it was, turned out to be no more than a shift in emphasis.

As was noted above, the nuclear reactor decisions demonstrated the limits of professional power, particularly the divided technical occupations claiming professional status within the nuclear policy system. This is largely due to the different nature of their role; that is, they are positioned within techno-bureaucracies and do not function on a one-to-one basis as do the legal and medical professions. The exceptions here being the people at the top of the hierarchy, such as Marshall, Hinton, Hill etc., who do have direct and influential access to Ministers and senior civil servants, but act really as a link to the elites of other groups in society. These are competing elites only in that they differ over details (like whether to adopt the AGR of PWR): their basic assumptions concerning the necessity for economic growth and secure energy supplies are shared. The technocrats generally influence policy through their domination of the hierarchies in which they are situated, making their organisations their own and leading to professionalised decision-making. They seek to use their position to define problems in terms that facilitate the wielding of professional power, arguing against the 'intervention' of politicians in their choices.

As was argued in chapter 5, this represents an essentially flawed view of democracy and the political process, but it is one that is fostered by technocratic belief in rational decision-making and scientific discovery. As was shown above, on the occasions when there have been expressions of discontent muttered audibly (particularly in the case of Hawkins, England and Lomer) early retirement has swiftly followed, underlining the tensions in the relationship between the political controllers and the public corporations, and underlining also the position of the corporations. The greatest influence of the technocrats has come from the need for them outline the options available to be pursued. Because of the technical ignorance of the politicians it has been necessary for the technocrats to formulate options which the politicians can then relate to the political needs that they perceive to exist. The standard of advice tendered by the AEA over the period of this study has ensured its replacement as the Government's chief nuclear adviser de facto if not yet de jure. The technical professionals then are influential, but only within the strict parameters allowed by the political structures of the United Kingdom: this means that advice and options presented to politicians must be tuned to the political goals of the system as a whole. Those goals led to the inevitable move from British gas reactors to American LWRs, a move that can be reversed only if the AGRs begin to show a substantial commercial advantage or the environmental concerns regarding the PWR become politically unanswerable. The dominance of the commercial ethos for the actions of the technocrats and how they have attempted to exploit it for professional goals of autonomy and expansion is explored in detail in the case studies which follow.

NOTES

1. There are no figures available however for an analysis of the actual numbers involved

8 The evolution of British Nuclear Fuels Limited

The previous four chapters have shown that although there are several aspects within the nuclear policy-system which tend to confirm the general propositions outlined in the Introduction, the actual policy process is far more complex than originally suggested there. Specifically, the first proposition, that technical professions provided an alternative communications system which could subvert political leadership, has been partly undermined by the realisation of the dominance of 'politics' in the British method of achieving a policy consensus. Thus, the other two propositions hypothesising a reduction in the political content of decision-making and that the decisions eventually obtained are the result of technical preference, have themselves to be qualified. The dialectic between the 'technical' and the 'political' provides the focus for much of the analysis which follows in these case studies. The first one, which forms the subject of this chapter, analyses the evolution of British Nuclear Fuels Limited (BNFL), a process which is an illustration of the effects upon Britain's nuclear decision-making framework of the organisational loyalty and the quest for occupational autonomy inherent to professional occupations. But many of the factors first outlined in the previous chapters are found in BNFL's development, particularly the need for technocrats to educate themselves politically if they are to obtain the (technical) decisions they seek from the politicians. This chapter traces the progress of that education and its utility for the management of BNFL during their quest for independence.

It is contended that the establishment of BNFL was largely due to the dual loyalty of the technocrats, as outlined above (Chapters 3-7) and noted by Dunleavy in his study of local politics (1980, p.111). The creation of a separate fuel company was a distinctive attempt to achieve occupational independence for the Production Group's technocrats. These people deliberately altered the functional parameters and the goals of the Production Group of the AEA in order to increase a need for their skills and facilitate a situation whereby they could command work autonomy. In order to aid the success of these aspirations the professional elite firmly set its goals within the ideological framework of the two major political parties. This served the technocrats well and by 1978 Con Allday, the Chairman and Chief Executive, could say that he found it:

> "...gratifying that both the major political parties in this country have committed themselves to support nuclear power and more recently to reprocessing in particular (1978, p.1)."

It can be argued that the political desires and ideological predispositions of the politicians and civil servants, combined with their technical ignorance, were exploited by the professional elite in their desire for increased autonomy, a situation paralleled in the JET project some years later. The analysis of this chapter follows these essential points, arguing that they illustrate the predominantly non-technocratic nature of the UK's policy-making procedures in even the most technically complex areas. The following sections first discuss the function of BNFL and the role of the technocrats therein, before analysing the nature of the dual loyalty found within the organisation. Finally there is a review of the background to the establishment of the company and an attempt to understand the reasoning behind the technical experts' movement towards autonomy, from the context of their position within the AEA in the 1960s and 1970s.

An Outline of BNFL

Established by statute in 1971, BNFL constitutes a monopoly of the nuclear fuel industry in Britain. By 1986 it employed over 16,000 people directly or in its two subsidiary and eight associated firms; it had assets of over £1,900m. and an annual turnover in excess of £629m., to which exports contributed at least £125m (Annual Report,1985-1986). The company manufactures fuel at its Northern factories for the ESI's nuclear power stations, and then reprocesses the spent fuel at its Sellafield (formerly Windscale) reprocessing plant in Cumbria. It also prepares the plutonium fuel for the UKAEA' fast reactor situated at Dounreay and stores excess fuel in readiness for a future fast reactor programme, although some of the plutonium produced at Sellafield is destined for a military purpose. In addition to these chores, the company produces a large range of intermediate products for the home and overseas markets, providing nuclear services to a wide range of countries thus:

> "The company's primary objective is to provide an efficient fuelling service to the nuclear power stations of the United Kingdom's Generating Boards and to meet the needs of other domestic customers. In addition the company seeks business overseas in order to achieve a greater measure of growth than that offered by the UK market alone, and to enable domestic customers to benefit from this expansion". (BNFL 11th Annual Report, p.11).

Inherent to these objectives are the commercial goals of a return on capital consistent with the legal, social and contractual framework within which the company operates and a 'high degree of self financing' (BNFL, 12th Annual Report, p.11), which in the latter half of the 1980s also means careful attention to safety, employee welfare, and relations with the local and wider communities (1985-1986, p.6).

It seems apparent that the company seeks to promote the impression that it operates within an ethos of hard-nosed commercialism, attempting to achieve a high degree of self-financing and performing within the market-place no differently from other business organisations. It is argued below that the commercial ethos is one that has been carefully constructed in order to facilitate the policy-making independence sought by the technocrats and consistent with the themes identified in Chapters 1 to 3, above.

Developments in the late 1980s, however, may well illustrate that this commercialising propensity has led to too high a profile for the company, which when combined with a series of leaks and accidents has aroused the wrath of environmentalists and Members of Parliament; the company being severely criticised for its operations at Sellafield by the Commons' Environment Committee (1986, Cmnd 191-1). Furthermore, the appointment of Christopher Harding in 1986, a non-technocrat from the Hanson Trust, appeared to be a deliberate attempt to reduce technocratic control and boost the commercial ethos. Whilst there is no likelihood of BNFL being privatised in the short-term, Harding's appointment and the changes subsequent to that, illustrated a deliberate shift towards the private-sector and the adoption of private-sector organisations and attitudes. These developments are symptomatic of the innovations brought by the Thatcher Governments to the public-sector, things such as the Raynor scrutiny, MINIS and the Financial Management Initiative, but they are also rooted deeper, their beginings nurtured by the factors that led to the development of BNFL.

The 1971 Atomic Energy Authority Act transferred the Authority's Trading Fund activities (set up in 1965) to two new companies: BNFL (formerly the Production Group of the UKAEA) and Amersham International (formerly the Radiochemical Centre). The Act was passed by the Conservative Government of Edward Heath during its 'Selsdon Man' phase, when it was anxious to reduce the direct intervention of the Government upon the economy and leave an unfettered industrial sector to achieve an export-led increase in economic growth (Stewart, 1978). The Conservative Bill, however, was closely modelled on the earlier version presented by Tony Benn, but which fell with the Labour Government in June 1970, and this attempted legislation was itself based upon a 1968 Ministry of Technology (Mintech) report following the 1967 recommendation by the House of Commons Select Committee on Science and Technology (SCST) which favoured such a move (HoC 381-1). This can also be traced back to even earlier recommendations from within the Authority itself (11th Annual Report of the AEA). The sections below which follow these steps on the path of the company's evolution illustrate several of the points first raised above (Chapter 1), namely that Gray and Jenkins's arguments concerning the important role of bureaucratic politics in securing policy objectives are applicable to this case-study (Gray and Jenkins, 1985, Chapter 2). The evolution of BNFL appears to actually have followed three of Weber's four classical paths to existence in that it was:

> "...created by one or more groups to carry out a function for which they perceived a specific need;
> split off from an existing organisation;
> and a group of people advocating a specific policy gained enough support to establish and operate an organisation devoted to that policy". (Downs, 1967, p.5).

Altogether five strands can be identified in the movement that established the company. First, the change in British Defence policy between 1958 and 1962, combined with the cut-back in the home-based Magnox power station programme in 1960. Together these two events led to large-scale lay-offs and cutbacks within the Authority. The third strand was an increasing scale of commercial links with Europe throughout the 1960s and, fourth, the setting-up of the separate Trading Fund within the Authority in 1965. Finally the Production Group itself began to establish a growing autonomy within the Authority and display a separate identity and esprit de corps that led to a desire for a separate company to grow within the ranks of the technocrats. This desire, it will be argued below, was shaped by the professional elite occupying high managerial positions; it was their ambition that provided the necessary momentum to secure the evolution of BNFL, a factor suggested by the analysis of Chapter 3 above.

Technocrats in BNFL

The dual loyalty of the technical professionals has been a consistent theme of this book since its identification in Chapter 2. The nature of the relationship of BNFL to its technocrats is similar to that of the Authority from which it evolved (see Chapter 4, above) and follows the pattern suggested by the analysis in Chapter 3. Within the UKAEA, the federal nature of the Authority led to the loyalty of the staff being focused upon the Group to which they belonged rather than the Authority per se: this developed into a sense of corporate identity even before the formation of BNFL. It is illustrated in the discussions of the following sections that the Production Group technocrats had their loyalty forged in the aftermath of the cutbacks in the early 1960s. As noted above, similar examples are found in other policy areas where there is a hierarchy of technocrats (Dunleavy, 1981, p.111) and in one sense the technical organisation in BNFL and the Production Group followed an almost Galbraithian pattern (1967), with groups of people trained in complementary skills being taken and welded into a single team devoted to organisational goals. As in the CEGB, personnel advanced in their careers through devotion to their field of expertise in the service of the company. BNFL, like the AEA, sees no conflict between professional loyalty and loyalty to the company. A former executive summed up the company's view by arguing that:

"There is no conflict between people identifying with their profession and with their employer. We encourage people to belong to their Institution. The postgraduate training for engineers has become, and will become, more organised through the institutions and we certainly cooperate as fully as we can".

He continued that in BNFL there was obviously 'nothing like the professional identification of doctors and lawyers' and that in the company, as in the AEA and CEGB the fact of being employed in a corporate structure meant that there were some engineers who identified quite closely with their professional institutions and others who 'couldn't care less'.

The policy-making structure of BNFL reflects its technical/AEA background and, like the Authority and the CEGB, its managers are predominantly technocrats. Typically for a commercial organisation, there are no reliable data detailing the professional breakdown of the management and work-force and, like private industry, there is no recognised dividing line between scientists and engineers. Senior management recognise that new recruits have a tendency to identify more with their professional sub-discipline than with the company and it is part of their personnel development to nurture a corporate identity. Certainly this is achieved by the time technocrats progress into senior managerial positions, a point expanded in wider organisational observations by theorists such as Downs (1967) in his study of bureaucracy, and Simon (1959) in his analysis of organisational decision-making. A former Board member of BNFL, argued:

"There is broadly a point at which we say to people 'You are now more of a manager than an engineer in this post",

a point that reiterates the practice of the AEA, promising technocrats being steered into appointments that carry an expanding managerial load. In BNFL care is taken to ensure a broad representative mix of the technical sub-disciplines at Board level, an appointment to which signals membership of both the professional and managerial elite.

Group Loyalty and Goal Transformation

In the 1960s the teams of BNFL's predecessor, the Production Group of the UKAEA, became the focus of the technocrats' organisational loyalty. Established in 1960, the Production Group appeared to have a secure future. A product of reorganisation due to the massive surge in defence-related work in the 1950s (New Scientist, vol.16, pp.141-3) the group was particularly busy at its huge diffusion plant at Capenhurst, where enriched uranium was produced for the weapons programme. The other sites in the North were kept busy by the preparations for the large Magnox programme ordered in the aftermath of the Suez crisis. A contemporary of Hinton's recalls that part of the reason for the wide range of the Production Group's mainly engineering functions was a frustration with the university style of the scientists at Harwell, an abstract approach to problems that reflected the contrast between pure science and engineering and led to the latter's exasperation with delays. Hinton, whose approach was determinedly applied, decided to establish his own production facilities at Risley and secure independence from Harwell rather than 'waste any more time battling' with the scientists. This manoeuvre reflected the arguments of those (Pitt and Smith, 1981; Gray and Jenkins, 1985) who emphasise the importance of understanding organisational motivation and bureaucratic politics in any study of public-sector policy-making. Hinton certainly exploited his own reputation, his organisation's resources, and the goals of the politicians, in order to expand his Group and achieve his aims.

The seeds of a nadir to blight this situation of rapid growth were sown even before the expansion had peaked. The 1958 Mutual Defence Agreement signed by the USA and Britain one year after the explosion of Britain's first H-bomb provided for the supply of British plutonium to America in return for enriched uranium (5th Annual Report of the UKAEA; The Guardian, 4/6/84). Apart from the strategic reason of not being totally reliant upon a foreign power for the raw material for nuclear weapons, the need for Capenhurst was negated by the agreement. Indeed, by 1962, collaboration between the two countries was so close that Kennedy bluntly told Macmillan that the UK did not need to possess an independent deterrent (Economist, 22/12/62).

A second blow was struck at the Production Group in the Government White Paper of June 1960, where pressure from the CEGB had succeeded in achieving a 're-phasing' of the Magnox programme (7th Annual Report of the AEA). In the event the re-phasing was a large cut-back in the projected work of the Group and the Authority commented drily that its commitments were 'considerably in excess' of current requirements (ibid, p.47). The thirteen per cent cut in the Authority budget that followed these developments led to over six thousand redundancies between 1961 and 1965, the heaviest blows falling on the Production Group with Capenhurst losing over half its staff. By 1964 several of the factories, Capenhurst in particular, were merely ticking over on a care and maintenance basis (8th, 9th and 10th Annual Reports of UKAEA). The senior staff of that era recalled it as 'a terrible time' and resolved to salvage their teams even as the Group was decimated. In 1964 John Hill became Member of the Board responsible for Production and this was to mark another turning point in the fortunes of the Group and provided an illustration of the method by which a professional/organisational elite obtained its goals.

In this period there was an Authority-wide central budgeting function, and capital expenditure above a certain level had to be approved by the Board; but despite this the different Groups functioned as though the Authority were a holding company. Hill and his senior managers determined to exploit the modicum of autonomy they possessed and to greatly expand it. The thrust of the assault upon the Production Group had been channelled through the Authority and after consolidation, the Group's managers began to consciously plan their bid for commercial freedom. The Group's technical expertise now transcended its defunct goals, and it used this expertise to construct a new role for the technocrats, transforming the goals of the Production Group in the process into a matrix that was both functionally desirable and allowed a functional independence. The professional ideology combined with organisational goals in the sense that there was a distinct move towards the 'preservation of significant levels of work autonomy for professional staffs' (Dunleavy, 1983, p.13).

But unlike Dunleavy's analysis of professionalism as interest corporatism, the technocrats did not simply argue that policy 'is just what professionals in the field do', since that was obsolete (ibid). The managerial/professional elite decided which of the skills they possessed could be utilised for functions that were commercially exploitable and would ensure the maximum amount of independence from both the AEA and the Government. They then systematically set about altering the parameters and ultimate goals of the employing organisation to reflect those skills and the products they were capable of supplying. Thus Dunleavy's observations can be firmed up with the argument that what the professionals did was to select goals that were achievable and guaranteed the greatest hope of work autonomy; then restructured their organisation's aims to reflect this. It was not a negative formulation of things they were already doing, but a positive attempt at a rational restructuring in order to secure their survival and expansion.

Hill and his deputy Franklin consciously opted to pursue a policy which they hoped would enable them to turn the fuel cycle venture of the Group into a commercial success. This afforded two benefits; first, it was believed that a commercial organisation run along conventional business lines enjoyed greater decisional independence than a bureaucratic sub-group of the Authority. Second, within the constructs of a liberal democracy it was felt that commercial success was a desirable yardstick by which to measure success per se and would win the Group external political allies. BNFL evolved as the vehicle for the occupational autonomy of the technical professions it employed, the senior staff of the Production Group resolved to pursue independence via several areas that they identified as requiring reform, and which if successfully achieved would aid them in their campaign. First, they felt there was excessive concern within the bureaucracy of the Authority upon the need to engage some Authority-wide services like health and safety research and development costs, items they perceived to be a needless duplication. Second, they felt a growing resentment at what was seen as the hampering of the practical revenue-earning potential of the Group through a set of rules and procedures that were a vestige of the pure research role of the Authority. A senior manager recalled that:

> "...Some of us were discontented about the arrangements. It did seem to us that if you were working in this organisation and you wanted money for some remote dream on research, you could get that money fairly easily. If you wanted it for something that would produce a direct economic benefit to your operations, then the criteria that were adopted were exactly those that would be adopted in an ordinary business".

It is clear that this was a situation in part exacerbated by the contrasting worldviews of scientists and engineers in the Authority.

This frustration led to intensive lobbying of the Board to remove many of the constraints Hill and others felt were stifling the fledgling commercial venture. Hill secured a major victory for the Production Group with the creation of the Trading Fund in 1965, a move which, accompanied by a general relaxation of Authority regulations, acknowledged much of the Group's case and prepared the ground for the realisation of its commercial potential (10th and 11th Annual Reports of the UKAEA; New Scientist vol.25, p.415). At Hill's insistence, the Fund was constructed to emulate the example established by the Post Office some years earlier. The Trading Fund effectively formed what The New Scientist called 'a new nationalised industry' (ibid). It provided for the Group to be 'allowed to retain all revenues' earned in the 'production and sale of reactor fuel elements, electricity, radio-isotopes and miscellaneous nuclear products' (10th Annual Report of the UKAEA, p.46). The Authority agreed that the factories of the Group were to be 'organised primarily to meet the commercial demands of the domestic nuclear power programme and overseas orders' (11th Annual Report of the UKAEA, p.9).

The new commercial orientation was reflected in a change in the titles of the higher management of the Group, a format geared towards the language of business: Hill became Managing Director, Franklin was named assistant Managing Director, and Dr. M. Davis was appointed to the newly-created post of Commercial Director. Apart from a very brief period when the Group flirted with the importation of commercial expertise from private industry, this latter post remained under the tutelage of the technocrats, with the chemist Con Allday occupying the position at the crucial time when BNFL was actually established. Allday went on to become Chairman and Chief Executive, being succeeded by Christopher Harding in 1986. Harding is a manager on loan from the Hanson Trust, he is not a technocrat, his appointment reflected a concern by the Government to inject even more of a commercial ethos into BNFL and represents (at least) something of a set-back for the technocrats.

It was the establishment of the Trading Fund which cemented the autonomy of the Production Group within the AEA and did much to consolidate the loyalty and group identity of the technocrats employed by it. A senior manager said that 'something of a contempt' developed 'for the rather lax financial standards, as we saw them, in other parts of the AEA'. A Chemical engineer employed by the Group during this period recalled that loyalty to the Group reached the stage where other parts of the AEA were regarded as dangerous rivals. An example, he argued, was a report he submitted favouring some work at Dounreay instead of within the Production Group network. His superior, he said, 'was furious. He told me Dounreay was in the Reactor Group and my responsibility was to the Production Group'. Identification with the employer was made easier for the employees because the Production Group was the Authority's engineering side; few scientists were employed in it and those who were tended to be of an applied nature. The rest of the AEA was largely unconcerned with these developments, being mostly staffed by scientists.

The other Groups tended to look to their own research and only a few minor objections were raised by those who felt that any loss of functions was a diminution of the Authority's power. But most members of the AEA did not have a view, and of those who did it was generally accepted that it seemed logical to hive off the engineers into a separate company as they were virtually autonomous in any case. Franklin, who was directly involved in the process of establishing BNFL, and was Allday's predecessor, argued that there was some minor opposition because of the reduction of the size of the Authority, but like other senior managers he argued most of the AEA did not have an opinion, preferring to concentrate on their own organisations. Of the senior managers involved, other than Hill, some were decidedly in favour 'because they had got the attitude that said 'You've got to convert things into businesses when you can' (interview with Franklin).

It is clear from their increasing orientation towards the world market, and also from their participation in international events, that the professional/managerial elite began to regard itself as a part of a wider international 'nuclear club'. Burn argues that in the 1950s and 1960s the predominant mood in the USA was towards a commercial, privatising movement, seeking to pass as much of the civil aspect of nuclear power as possible into the private sector (Burn, 1978, Part 1). It can be argued that this had a propensity to influence the managers in the old Industrial and Production Groups who were developing a fuel cycle process ripe for commercial exploitation and free from the fetters of being an extension of a state-run hierarchy. A small policy directorate meeting in the mid-1950s under J.C.C. Stewart (later member for reactors) and including Hill and Franklin, was certainly toying with a move towards privatisation. Franklin recalled:

> "It was that group that really served as the spur for the commercialisation of attitudes, and almost exclusively that group".

By the time the collapse in demand for civil and military products occurred, there already existed a small group of very powerful people in influential positions convinced of the need to transcend the traditional constraints of the AEA. As with so much of the policy-making in British governmental and quasi-governmental circles, the role of individuals at certain key periods of decision-making was crucial. The events of 1960-1965 provided the catalyst to implement the various ideas mooted during the discussions of Stewart's directorate.

Construction of a Wider Base

The determined expansion into overseas markets and the development of new products to serve those markets was an attempt by the professional elite to move away from a reliance upon a monopoly customer in the form of the CEGB and the dictates of the AEA. It was, therefore, an important part of the process towards the attainment of greater autonomy as it allowed the claims to autonomous status to be validated by measurable results. Where direct market penetration was prevented by foreign governments the Group sought to establish joint companies with a country's indigenous organisations. Hill went to great lengths to link the Production Group to Europe (see above Chapter 4) and encouraged his higher management to do likewise. This was partly because of the protectionism of the American market and partly because of the belief by Hill that the future for British technology lay in joint ventures with the members of the EEC. One of the managers recalled the policy as an attempt:

"...to see whether we could sell internationally those sorts of service that were in the nuclear fuel cycle, but were not specific to Magnox fuel. We tried to see whether we could develop markets in those products which were not reactor specific".

It tried to pick out those elements of their expertise that had a greater universality of application and could be sold. This is in contrast to the usual practice of professions, particularly those established social professions such as Law and Medicine, where it can be argued that professional power is obtained from their ability to define their client's needs for him (Wilding, 1982; Larson, 1977). The behaviour of the Production Group's professional/managerial elite tends to confirm the subservient nature of the technocrats, as argued in Chapters 2 and 3, the goals of the professional ideology being attained through the attempts at organisational domination, and not from a position of social status. The technocrats of the Group could not define their clients' needs for them but had to actively seek out clients by first defining the type of services they as a group could provide and then embarking on market research to establish a need. The research also allowed them to trim their services to meet expressed demands. The links of the technical professions with the commercial ethos were never clearer than in this exercise; it illustrated the narrow limits to technocratic influence that is independent of political allies whilst simultaneously showing that the path towards organisational autonomy lay within the commercial, not the professional, ethos.

An important element in this was the dawning realisation that there were not going to be large numbers of export orders for the British Magnox reactors. The Economist called the Production Group's efforts to make good the gaps in orders left by this 'superhuman' (Economist, 4/5/63). The efforts began to pay off and soon the teams experienced expansion, the factories at Capenhurst and Springfields working up gradually to commercial production (New Scientist, vol.19, p.337). Links were forged with most of Western Europe's nuclear-based companies, as well as close trading ties with the Far East. The two most politically important associations attained were concerned with reprocessing spent nuclear fuel (between Britain, France and Japan) and the enrichment of uranium to be used in Britain AGR and the American light-water reactors (between Britain, Germany and the Netherlands under the 1970 Treaty of Almalo). Attempts to supply enriched uranium to the USA were thwarted by American fears about foreign domination of their strategic industries; they saw it as a British move to 'capture the world markets' and off-load spare capacity -something they had hoped to achieve themselves because of their own enormous over-capacity (New Scientist. vol.19, p.337;

vol.37, p.356). A perception began to emerge in Europe that America was a major threat to European independence in the nuclear industry, and a desire to form European cartels gathered pace, aiding the Production Group in its attempts to link up with EEC companies (ibid). A string of important contracts were signed between 1965 and 1975 and BNFL became firmly enmeshed with the European nuclear industry (Annual Reports of the UKAEA, 9-16; Atom, Nov. 1963; Jan. 1965; 1966; 1967; 1968; 1969-72; Guardian, 4/6/84).

It can be argued, therefore, that the commercial enterprise of the professional/managerial elite was a dominant factor in the evolution of BNFL from within the Production Group of the AEA. The majority of business contacts sought were negotiated by the technocrats themselves. As the dominant factor in the organisation, leading to a professionalised system, the Production Group had its goals motivated by the de facto influences of the ideology of 'professionalism'. Unlike the social professions, however, this encompassed not a rejection of the ideals of the commercial market-place (Larson, 1977; Johnson, 1972) but an embracing of those goals as the surest route to organisational autonomy. This is not to argue that the managerial elite thought in explicitly 'professional' terms; indeed, they rarely viewed their problems and goals in that way, in contrast to the legal and medical professions. But the practical expression of their professional formation and the fusion (in a state of corporate patronage) of their professional and organisational allegiance, led in this instance to their organisational behaviour being directed by aspects of collective behaviour normally associated with the ideology of 'professionalism'. The professional elite, in the pursuance of its objectives, found it necessary to forge some strategic links with powerful political allies, and it is this aspect which is now examined.

Strategic Links

To argue that the technocrats sought strategic links with the ideologies of the two major political parties is not to accept the implications of Wilding's analysis which effectively portrays the professions as a pressure grouping (1982). Neither does it accede to the view of the functionalists which argues that they fulfil a necessary social function; or that of Durkheim who views the professions as a benign and essential check upon the logical application of the capitalist market (1966). It is contended that as the product of essential elements in industrial capitalism, it is logical to assume that the technocrats seek to emulate those aspects of a liberal-democratic construct that afford the greatest hope of achieving occupational autonomy and control over their own market. A statist organisation that is a sub-group of the AEA, although technocrat-dominated, did not fulfil that function. The Commercial ethos, a strategic necessity , was linked to an appeal to the predispositions of the politicians and civil servants, groups that were the effective policy-arbiters in the nuclear power policy-system.

The commercial strategy appealed to the ideological prejudices of Conservative (and many Labour) politicians. To form a nuclear fuel company helped to de-mystify the aura surrounding nuclear energy, whilst the full order book held out the promise of large returns on investment. For this reason Conservative politicians initially sought a major private-sector shareholding in the company, although Labour wished to restrict it to a minority (Atom, June 1970). By 1971 the Conservatives had also been converted to the idea of retaining a majority public shareholding (Atom, January 1971) despite the first attempt at forming the company via the necessary Act of Parliament, falling in the Lords due to Conservative insistence upon the dominance of the private sector. The four factors which finally convinced the Conservative Government of Edward Heath to initially retain a total public shareholding were: outright military secrets from the Group's work on atomic weapons' materials; possible military secrets being extrapolated from the civil work, such was the relationship of the two; the need to maintain staff conditions and terms of employment; and the sheer cost of investing in the company without any significant short-term gains (Atom, June 1970) [25]. In essence, however, there had been general bi-partisan agreement upon the need for a hived-off fuel company from the time it was first recommended by the Commons Select Committee on Science and Technology in 1967 (Recommendation no.3). Tam Dalyell wrote:

'Whitehall is sure that for once it has spawned a company with excellent prospects for showing a handsome return for the taxpayer' (New Scientist vol.50, p.2).

The hiving-off reflected the prevailing managerialist mood of politicians, a desire to achieve efficiency and effectiveness in general (Stewart, 1978).

The professional elite of the Production Group sought to promote the belief that commercial independence conformed with the goals of the Government. It was a shrewd move by the senior management to adopt the trappings of commercial business as this helped initially to break down the fear and suspicion surrounding the subject of nuclear power (fear since rekindled by the Chernobyl accident and by a series of leaks at the company's Sellafield plant). It was the aura of fear that did much to facilitate political control, though technically ignorant politicians tended to leave matters of safety to the experts. But any move to shed the mystique aided greater independence from political interference and an image as 'just another company' was useful for this purpose. It was a move that certainly attracted some criticism from others in the nuclear industry, CEGB's senior management arguing that BNFL's management occasionally allowed the commercial ethos to conflict with their technical role, a charge vehemently denied by BNFL. Yet there still exists a view that the fuel cycle of nuclear reactors is 'one of those things which are too sensitive to run as a commercial undertaking', according to one CEGB executive.

There were other, stronger, views expressed in the CEGB and elsewhere which cannot, for obvious reasons, be quoted here, but which illustrate a deep distrust of BNFL and Government moves to emphasise the profit motive. This antipathy is perhaps symptomatic of a wider public-sector ethos of service and not aimed at BNFL per se. The view of many in BNFL, that there is no real difference between the private and public sectors reflected the growing commercialisation of the public sector that paved the way for privatisation and also served to reduce the barriers between the two sectors, one of the effects of a growth in commercially-orientated quasi-governmental agencies (Greenwood and Wilson, 1984, Chapter 10).

The main propaganda targets of the commercial drive were the politicians and civil servants, and they were suitably impressed. A seductive argument enunciated by the Production Group to the politicians was the twin promise that a commercially independent company would be able to reduce fuel costs and fund this with an export market that promised lucrative profits. It was calculated that in any power station fuel could account for up to a third of the cost of electricity sent out (New Scientist, vol.43). However robust the statistics, it was clear that any country that could reduce its power costs would increase its competitive advantage, having this funded by exports was a double bonus, both for the balance of payments and the electricity consumer. For governments wrestling with rising fuel costs and problems with overseas trade deficits, the promises of the technocrats were very appealing. As well as the close connections with Europe, Hill and Franklin had built up a series of profitable contracts with Japan, the sale of the Magnox at Tokai Mura supplying the initial opening. Wynne (1982) provides a detailed account and analysis of the Japanese connection and how it led to the building of the Thermal Oxide Reprocessing Plant (THORP) at Windscale (see below, Chapter 9). In the late 1960s and early 1970s the fears of the environmentalist lobby were still either muted or inchoate, and the promise of substantial revenue from reprocessing Japanese fuel was held to be a major political coup for the technocrats, particularly as it was achieved in the face of stiff opposition from France and America. The Production Group (and then BNFL) promised to reprocess spent fuel as part of a unique package deal, unique at that time because Britain possessed the only integrated fuel-cycle system in the world. Other countries had followed the American lead in allowing the reactor manufacturers to also make the fuel elements. Britain, however, could provide the type of customer-orientated specialist facilities sought by utilities with a nuclear capacity. The technocrats, therefore, firmly linked their goals to those of the politicians, arguing that hiving off the fuel cycle would substantially contribute to the economic wellbeing of the country: a ruse identified elsewhere in the nuclear industry (see above, Chapters 4 to 7), albeit in a moderated form.

It can be argued that one of the important political motives behind the establishment of BNFL was of a purely punitive nature: a desire to reduce the power of the UKAEA. The cavalier attitude of the Authority in previous years with regard to the norms of Whitehall and in particular to costs, had earned it many black marks, and it failed to redeem itself by tendering advice that yielded poor results, as the analysis of the previous chapters has attempted to illustrate. A senior member of the SCST was clear that such a motive lay behind his decision to support the 1967 recommendation to split up the Authority:

"The AEA had got too big and was trying to do things it had no remit for. It should be primarily a research organisation, that's what it does well. The fuel reprocessing aspect is an engineering job and we felt it was best to separate it and give it to engineers".

Whilst a senior member of the Production Group felt that the establishment of the company reflected a settling of old scores by the civil service, the Treasury in particular:

"It's not beyond belief that ministers were advised in that direction by civil servants on the grounds that they felt it would be no bad thing if the Authority were diminished, and they could see that as a way of doing it. The AEA in its early days had been able, through the power of Lord Plowden and the direct support of Prime Ministers, to take a robust line with everybody, including the Treasury. One's experience is that this is not easily forgiven in Whitehall".

A view which echoes many of the arguments contained in Gray and Jenkins's model of bureaucratic politics (1985) and the discussions of Chapters 4 to 7 above.

In the sense that the professional elite saw it as part of their job to hive out those parts of the Authority that were clearly no longer research-based, the policy initiatives for BNFL's evolution were internal and apolitical. But this move coincided with a political momentum that had similar goals for different reasons, a momentum exploited by the Production Group's managers, further aided by the move of John Hill to the Chairmanship of the UKAEA.

A factor that earned the managers of the Production Group some considerable political favour was the network of painstakingly negotiated connections with Europe, as the extent of these became clear at a time when Britain was attempting to join the EEC. In February 1969 the European Heads of State reiterated previous statements telling Britain that if it wished to accede to the Community then it had to increase collaboration in European projects (Economist, 15/2/69). One particular project which the Europeans were keen to press was the controversial (and still heavily classified) Anglo-Dutch-German project on the gas centrifuge method of gaining enriched uranium. The groundwork for this effort had been carefully laid by the Production Group technocrats and it was to achieve fruition in the Treaty of Almalo, an event that was steeped in more political significance than technical importance and was to earn the Group the lasting favour of both Conservative and Labour politicians. An added impetus was injected into the negotiations when the Soviet Union expressed concern that Germany should be involved in such a programme, since it was based upon the technology from which H-bombs were manufactured. The USSR feared that it was in contravention of the 1954 principles which established the Federal Republic (Nature, vol.221, p.985). The other nations of Western Europe were also anxious that Britain should join the project as a counter to German power (ibid).

The Treaty, although a hard-headed commercial decision, was an act of political restraint; it was a complex political decision. Professional influence was an integral part of the policy process leading up to the Treaty: the technocrats were essential in interpreting and explaining the implications of the technical developments to the policy-arbiters and in conducting the preliminary negotiations. Although it was a political decision, the minutiae of the Treaty were largely defined by the technocrats.

It can be seen, therefore, that a series of market-orientated moves begun in response to Government cut-backs had resulted in a series of politically significant ventures. One senior BNFL official, commenting on Almalo, argued:

"The Government and the AEA encouraged us to get involved for two reasons: first, that of security, it was felt that it was necessary to keep closely in touch with any developments in this field. Second, there was an obvious motive of keeping on top of technical advances".

As early as 1966 the politicians had defined the importance of the project for them, a definition almost entirely informed by technocratic interpretation:

"It was an area with a large export potential for the UK: there was an obvious disadvantage in being dependent on foreign countries; it was an opportunity to reduce the balance of payments deficit; it provided another opportunity to keep British technology to the forefront of world technology" (Atom, January 1966, quoting Frank Cousins, the Minister responsible).

It was the successful identification of their technical goals with these and other political aims that allowed the technocrats of the Production Group to transform their organisation in 1971 into British Nuclear Fuels Limited.

CONCLUSION

An understanding of the role of the ideology of professionalism (see Chapters 2 and 3) as it relates to the technical professions is a useful means of giving a valid insight into the organisational motivation and behaviour that led to the evolution of BNFL. As in previous chapters, this case study has attempted to show that professional motivation in the corporate sector of the economy is closely linked to organisational motivation, a result of the dual loyalty inherent to the technocrats. In BNFL, as in the CEGB and UKAEA, the professional and managerial elites of the sub-disciplines overlapped considerably. It can be argued that the political decisions taken with regard to the activity and structure of the Production Group and BNFL, can only be fully appreciated when cognizance is taken of the way in which these decisions were influenced, or even structured by the organisational expression of the professional ideology and its goals. Examples found in this chapter include the establishment of the Trading Fund; the European collaboration; the integrated fuel cycle; the decision to massively expand reprocessing facilities and (of course) the decision to establish BNFL per se. Certainly there were other more political factors involved, as noted above: a general desire to reduce the role of the AEA and a desire to foment European integration being but two out of a welter of often complex and conflicting economic, strategic and technical goals.

But a unifying thread amongst all these factors was the role of the technocrats themselves and their desire to attain the professional/organisational goals of occupational autonomy and control of their work content. The analysis of the dialectic between the technical and the political continues below with a discussion of the Windscale Inquiry, a period when the political and technical elite went to considerable efforts to define what had plainly become a political issue, in technical terms.

9 The Windscale Inquiry

In this chapter the analysis is concerned to explore the implications of BNFL's commercial ethos in more detail. Nowhere were these more apparent than at the Windscale Inquiry and this is why that event provides the second case study. BNFL's commercial pretensions are the subject of further review because as the last chapter attempted to show, it was through its central and dominant concern to achieve commercial success that the company sought to gain political favour and organisational (and thereby also professional) autonomy. Thus, by probing its pursuit of a commercial foundation it is possible to expand the understanding of those other motivations central to this book, namely the way technocrats are involved in framing and implementing policy to advance their own goals. After outlining the background to the Inquiry, this chapter will proceed to discuss the contradictions thrown up by technocrats engaged upon a commercial venture claiming immunity from lay attack due to their professional competence. A subject germane to an understanding of the technocratic role in the Inquiry, the focus of the section which follows it, and leading to an appraisal of the lessons Windscale provides for understanding technocratic decision making.

Discussion associated with the Windscale Inquiry is informed by the technocrats' view of politics. Earlier chapters have dealt at length with what could be seen as an essentially flawed view of democracy, a belief in the correctness of technocracy to make decisions which have important social consequences (Borguslaw and Ellul, 1981). A constitutionalist view of Parliament and politicians, which argues that they cannot be expected to be technically expert because it is their role to be politically inquisitive and operate on that level effectively (Johnson, 1977; 1981), appears to carry little weight within the serried ranks of the technocrats. Indeed it is explicitly criticised, Nature arguing that 'the British Democratic Tradition had gone wild' in its anxious attempt to secure 'public debate over nuclear power' (vol.269, pp.640-641) during the wider arguments about Windscale's expansion. A view the marathon Sizewell Inquiry confirmed in both technocrats and the Conservative Government.

The approach of this chapter has been adopted partly because of the nature of the content of the interviews undertaken for its research. Unfortunately the majority of those interviewed stressed their desire that no quotes, specific views, or even general observations be attributed to them. Several important points are apparent, however: the Inquiry represented a paradox for BNFL's technocrats, the very act of securing greater autonomy bringing calls from observers for the company to become more accountable; the nature of technical professionalism is again exposed; the technical/political dialectic betrays the links of 'professionalism', 'scientism' and 'rationality'. The first of these points is the one that caused greatest angst within the policy-system. The policy of commercialisation strayed from the world of official secrets and technical rationality into the spotlight of economic desirability, awakening a Pandora's box of slumbering daemons that united to harass the technocrats from then until the present. Specifically the civil and military uses of nuclear power became linked together for the first time; the need for security against theft of nuclear material by terrorists; safety and environmental impact, all joined at the same time to subject the entire nuclear programme to a fundamental and continuing process of questioning. The other points, although incidental to the Inquiry itself, are substantive to this study. Windscale provided an illustration of the lack of professional cohesion within the technocratic occupations and the primacy of their loyalty to the employing organisation. Furthermore, it showed that the political elite was largely dependent upon technical advice and it actually preferred issues to be presented as technical decisions because this extracted politicians from the mire of debates about things they had little interest in, but which could be electorally damaging. As previous chapters have attempted to show, the politicians had ample scope for political intervention: what they actually sought quite often was a simple choice.

A desire to avoid the politicisation of nuclear power was the governing motive of politicians from both political parties from the early years of the programme (Williams, 1980, introduction). Whilst he was in office, this also applied to Tony Benn, although as is shown below, his behaviour in the Windscale affair was initially aimed at increasing public information; it is since leaving office that he has become an opponent of nuclear power (Benn, 1984, evidence to Sizewell Inquiry). The general political attitude reflected a bias in favour of crisis management and a fervent wish to avoid conflict over issues that it was perceived carried little or no electoral bonus; similarities with other 'depoliticised' or 'depoliticising' spheres like the NHS and full-employment commitments, are obvious (Smith, 1979; Middlemas, 1979). This technocratic ideology, inherent to corporatising processes (see above, Chapter 2) transmitted itself throughout the public (and private) sectors. An example was a meeting held at the Department of Energy (DEn) on 8th July 1977, between UKAEA officials and DEn civil servants to co-ordinate their approach to the Inquiry (despite the Department's role as an independent witness). In a discussion about the long-term storage of irradiated fuel, an important issue of the Inquiry, the Authority duly noted that:

> "We hoped that questioning on this matter would be confined to technical issues, though clearly the enquiry would wish to know why the subject had not been more fully investigated in the past" [1].

In the light of experience this was to prove a perspicacious comment.

The Background to Windscale

The Windscale Local Planning Inquiry was established by Labour's Peter Shore who used his powers as Secretary of State at the Department of the Environment (DOE) under Section 35 of the 1971 Planning Act, to enquire into a proposal by BNFL to expand part of their Windscale (now Sellafield) site. The company sought permission to build a Thermal Oxide Reprocessing Plant (THORP) and the Inquiry, chaired by Mr. Justice (now Lord Justice) Sir Roger Parker, ran for one hundred days, from 14 June to 3 November 1977. Parker was aided by two eminent technical assessors, Sir Edward Pochin, a radiobiologist, and Sir Frederick Warner, an engineer who had worked with Parker during his investigations into the Flixborough disaster. Parker was charged with reporting to Shore a series of recommendations with regard to the desirability of the THORP proposal.

Designed to reprocess spent British and foreign nuclear fuel, THORP was (and is) an essential element in BNFL's commercial drive. The expansion was seen as a prerequisite to capturing the large Japanese markets sought by the company, and similarly it was viewed by the British nuclear industry as integral to the UK's progression towards a commercial programme of Fast Breeder Reactors (FBRs), the prototype of which was situated at Dounraey. The Japanese contract alone was worth at least £400m to BNFL in current (1975) prices (The Economist, 25/10/75, p.90), an alluring carrot with which to tempt the politicians into granting permission. To this was added the tantalising promise of an advance payment by Japan, totalling £150m, half of the 1975 cost of building the plant. Furthermore, the Japanese promised to accept the return of their reprocessed products, undertook to pay any unforeseen cost increases, and agreed to the right of BNFL to renege in total on the agreement if technology could not be perfected to fix waste in glass (ibid). These generous conditions obviously facilitated Government approval. In a national climate of currency and energy crises, IMF loans and stringent financial controls on the economy, BNFL had landed a deal by which a major industrial competitor would pay in advance a large part of the development and construction costs of a massive and technologically unique industrial complex.

Much of the alacrity displayed by the Japanese was due to the uncertain world reprocessing situation. A general escalation in costs due to unforeseen technical difficulties, occurred simultaneously with an American moratorium on reprocessing because of its association with weapons production; a factor that helped lead to unfavourable domestic opinion in Japan itself. Japan's lack of raw resources combined with the oil crises of the early 1970-s and the other factors outlined, spurred that country into its talks with BNFL (Wynne, 1982, pp.42-44). It was, however, those same factors which led the Daily Mirror (21/10/75), other sections of the media and the growing opposition, to cast Britain in the role of the world's 'nuclear dustbin', "subjecting herself to the ultimate indignity of accepting bribes to take foreign waste" (Wynne, op.cit.). BNFL dismissed such qualms and publicised the commercial benefits: Franklin argued that the company could earn £100m a year throughout the 1980s if THORP were built (New Scientist, vol.66, no.966, p.710). THORP would more than double production and have the added benefit, according to Allday, of helping to conserve the world's energy resources; it was an environmentally and economically acceptable option, he argued (Atom, 7/79, pp.136-139). The company saw THORP as a natural, incremental progression in the policy embarked upon by the Production Group a decade earlier. By March 1976, the Minister responsible for BNFL (Tony Benn) agreed with this argument, and gave permission for the Japanese contract to proceed on the understanding that THORP would be built.

This also signalled a further step on the way to an FBR programme, something Allday and others in the policy-system felt was an inevitable facet of the British nuclear development. In identifying THORP as a prerequisite to the FBR programme, Sweet argued, was one of the reasons why the opponents were so anxious to prevent its construction (1980, pp.1-3). The conservationist lobby were fearful that such a programme would 'dot reprocessing plants' all over the UK, instead of investing resources in alternative renewable energy sources (ibid, and pp.23-31). The unexpected decision taken by Shore to hold a public inquiry came as a shock to the technocrats who viewed THORP as an entirely logical move within the policy parameters that had evolved (Economist, 27/11/77, pp.99-100; New Scientist, no.1029, p.509). It was the type of political situation both they and the DEn had tried to avoid and the fact that opposition groups had led to the DoE outflanking them and the Cabinet goes some way towards corroborating the arguments of those who view Whitehall in terms of a federal structure of competing interests (Jones, 1980; Ashford, 1981 a,b; Gray and Jenkins, 1985). Certainly it was an embarrassing manoeuvre for the Government as a whole, since approval in principle had been given and it was now clear that this policy was to be tested at a public inquiry.

The Inquiry generated a large corpus devoted to several aspects of the issues involved. Parker's ninety-page Report (1978) in favour of THORP was based upon over 8,000 pages of oral transcript and about a million pieces of photocopy (Economist, 13/5/78, p.25). Apart from The Guardian's descriptive paperback (1978), the initial critiques of the Inquiry were issued from the perspective of the environmentalist opposition and criticised both the Inspector's decision and his reasons for it. Patterson and Conroy (1978) not only saw THORP as an environmental menace, but argued that Parker's Report was 'insufferably patronising and inexcusably slipshod' (ibid, Chapter 3). Conroy also attacked the inherent suppositions of the nuclear industry the Inquiry revealed, arguing that a reliance upon nuclear power was uneconomical and fraught with safety problems; he called instead for a new approach along the 'soft' paths of renewable energy (1978b). This reflected the outright rejectionist stance of the Green Movement, an holistic approach rejecting much of industrial society per se.

Pearce et al (1979) and Williams (1980) were more concerned to explore the Inquiry from a perspective analysing its relevance to democratic procedures for deciding policy. The former concentrated upon the efficacy of formal planning procedures and the latter more upon the issues of accountability involved. Williams saw the case as having political, economic, social, scientific and technological components and repercussions (1980, Chapters 11 and 12) with the Inspector viewing BNFL as the responsible public body that he presumed to be acting in the public interest (ibid, pp.315-320). It would have been remarkable, argued Williams, if Parker had made any other decision; he did what he was expected to in approving the application; it was a political decision. This is not to call the affair a charade: Parker was 'fully convinced of the merits of BNFL's proposal' (ibid). But the Inquiry was not an example of public participation (much less public accountability) in the policy process; Williams argued that it was merely consultation, with the Report itself being there only to inform the Minister.

Wynne pursues these points from a radical perspective (1982). He viewed the Inquiry as an elaborate and necessary ritual aimed at securing public consent to the technocratic/political goals. He argued that the ideology of 'scientism' was used to deny the validity of values and subjective argument (see above, Chapter 2, for the links with 'professionalism'). There are similarities here with the arguments of Habermas (1979) and even Ellul and Borguslaw (1981), namely that the growth of technocracy serves to depoliticise, and therefore disenfranchise the wider populace. Windscale was 'a major ceremony' which 'rededicated the faith' at 'a time of crisis' but in a reductionist manner that sought to ensure a limited access to the policy system (Wynne, 1982, p.161-173). It was, Wynne argued, an attempt to maintain the public in the role of a political peasantry (ibid), a view that has remarkable similarities with the work of Illich (1977) if the activities of the technocrats are viewed from the approach of the ideology of 'professionalism'. Clearly the nuclear industry and the Government were concerned to limit considerations as much as possible to those of a technical nature (see above, footnote [1]), and opponents were viewed as being ill-informed or hopelessly subjective. But it was precisely this organisational/professional approach to policy-making that led to nuclear opposition groups forming and raised the questioning of nuclear power and technocratic control over it.

THE CONTRADICTION OF WINDSCALE

BNFL's adaptation from a bureaucratic sub-group into a commercially aggressive corporation had been so successful that a growing concern emerged about the company's lack of political accountability (Wynne, 1982, Chapters 3 and 5). Groups including the Town and Country Planning Association (TCPA) and Friends of the Earth (FOE) lobbied for an Inquiry into THORP not only because they were opposed to the scheme, but because they were also anxious that the company had made too little information available for a proper review of its future proposals (New Scientist, vol.72, n.1024, p.198). BNFL's management had secured autonomy at the expense of attention to good public relations, and that very autonomy was itself now to be subjected to public criticism. By claiming commercial as well as technical expertise, the technocrats had shifted the basis for their claims to autonomy and laid themselves open to critical analyses based upon the economic worth of their product. The result was a questioning of both their commercial and technical ability; furthermore, the two claims could appear to be in conflict with each other, and it was argued in some parts of the policy-system that the emphasis upon commercial viability was at the expense of commercial excellence (see above (Chapter 8). Commercial goals are more open to criticism and alternative interpretation by non-experts than are technical and professional goals; thus BNFL's linkage of the two opened the company's strategy to non-technical critiques of the sort that occurred at the Inquiry (Wynne, 1982; Stott and Taylor, 1980).

Commercial goals in BNFL began to pre-set technical goals, undermining a foundation of the professional ideology; that work autonomy is necessary due to the control of a monopoly of knowledge exercised in the public good. Whilst such a situation probably never fully existed in technical professionalism the desire to increase greatly the reprocessing facilities at Windscale was based almost entirely upon commercial rather than technical grounds and the company thereby relinquished a major tenet of the technocratic faith. It became harder, therefore, for the nuclear industry simply to dismiss objectors for their lack of objectivity, even though that was the usual charge levelled by BNFL and the AEA during their publicity campaigns in favour of THORP between 1974 and 1977 (Atom, no.217, pp.255-6) [5]. But a charge of 'wishful thinking' in technical terms could have been applied by the objectors to BNFL since much of the detailed design work (and its concomitant problems) remained to be done by the engineers, even as plans for the Inquiry were being laid. This latter charge was a point partially conceded by Hill. In most of their comments, however, the technocrats blandly asserted that 'radiation will not cause problems to the public at large' and 'neither will the problem of radioactive waste be serious, again as far as the public is concerned' (ibid) because resources were being devoted on a major scale to solving them. In essence the technocrats were arguing that decisions should be left to them because their technical expertise had always triumphed over adversity in the past; a subjective viewpoint which to the opponents increasingly appeared to lack credibility.

The primacy of the commercial over the technical within BNFL appeared even more entrenched when after a 1974 accident Hill said that the problems of scaling-up facilities to build THORP were more difficult than had previously been allowed for (Atom, ibid. pp.255-6, 231, p.9), although the company asserted that they had learned from their mistakes (ibid). Yet, despite their move away from their technical foundations in terms of reasoning for their actions, the company still used the claim to superior knowledge as a basis for their right to autonomy in management; attacking 'sensation-seeking' politicians and journalists, Hill argued that the nuclear opponents had:

> "No real knowledge or interest in the subject and (used) emotion as their main instrument". (Atom, no.257, p.54)

Franklin argued that the acceptance of technical difficulties was itself a commercial ruse to 'talk up' the price of reprocessing during and after the Paris conference of 1975, in order to secure even better terms in subsequent foreign orders (Author's interview). Increased safety requirements, however, did lead to a rise in costs and the company sought to offset these through higher prices in overseas contracts (New Scientist, vol.60, no.955, p.710).

At the commencement of the Inquiry, BNFL's case, therefore, rested upon commercial arguments which were presented in the form of sixteen points by their advocate, Lord Silsoe. They specifically rejected the notion the THORP depended upon the FBR for its existence, quite the contrary, but underlined the commercial benefits accruing from overseas trade and the conservation of domestic stocks of uranium via reprocessing (Stott and Taylor, 1980, p.20-33). The company also pointed to the 'dire' consequences if THORP were refused, arguing that the local economy of Cumbria would be adversely affected, in contrast to the much-needed expansion in employment should the plant be constructed. In favour of THORP and lined alongside BNFL (albeit occasionally at a discreet distance) were the ESI, the AEA, some Government departments and the local labour movement which included the local authorities. In opposition were over twenty groups which ranged from the Isle of Man Government, the TCPA, FOE, the Quakers and British Council of Churches, through to some of the more bizarre fringe elements of the Green Movement (ibid; Wynne, 1982). The opposition coalesced around the issues noted above, namely the links between civil and military uses of nuclear power; the environmental and safety aspects; the large costs involved in developing THORP compared to renewable sources of energy and conservation. A debate rehearsed in even more detail at the Sizewell Inquiry between 1982 and 1985.

The burgeoning opposition to THORP had, before the Inquiry, led Benn to open a series of public debates in Cumbria and London on the subject. By March of 1976, feeling he had done as much as possible to further public awareness (Benn, 1984, op.cit.) he gave permission for THORP to proceed. Immediately the opponents lobbied the DoE for an Inquiry, and the local authorities, also under severe pressure, referred the matter to Shore, who in December announced the decision to hold an inquiry. Initially he had resisted the pressure, but violent civil disorder on the continent against nuclear expansion had convinced him of the need to defuse the British situation (Wynne, 1982). BNFL were disappointed; Allday argued that the company had thought 'the decision had been made' (Economist, 27/11/76) and the inquiry option was 'tragic' as it would lead to a delay which could lose the valuable Japanese contracts (ibid, 18/12/76, p.15). The technocrats therefore felt let down by the Government; there was a widespread view that the process was a time-wasting gesture that could have had serious repercussions.

Technocrats and the Inquiry

The literature has analysed Parker's emphatic findings in some considerable detail (Wynne, 1982; Stott and Taylor, 1980; Williams, 1980; Pearce et al, 1979). On all the points raised by BNFL and the objectors, Parker was decidedly on the side of BNFL and rejected any arguments to the contrary. As noted above, the opponents were outraged at what they perceived was shoddy treatment (Patterson and Conroy, 1978) in the Report, but in his role as an arbiter it can be argued that Parker had little choice but to opt for the stance he did. There are two possible reasons for this: the role of the technical assessors; and the nature of the professional disunity at the Inquiry.

Although an eminent lawyer, Parker was obviously a layman with respect to the finer details of the nuclear industry. Despite an evident ability to read into a subject, the Inspector relied upon his two expert assessors to interpret much of the technical data set before him. During the Inquiry it appears that they assisted Parker in three ways. First, they would analyse and comment upon proofs of evidence the day before they were due to be read out in order to alert Parker to any flaws or sections that required further examination. Second, the assessors were authorised to ask questions of the witnesses in order to clarify a point or expose an obvious error of fact. Finally they would discuss with Parker aspects of the evidence after the witnesses had been cross-examined. The writing of the Report was based upon the evidence thus analysed and early drafts were sent by Parker to the assessors for their comments. Both assessors fully supported Parker's final draft, and once the Inspector had included their views in his Report they were his responsibility (Pearce et al, 1979, pp.93-97). The importance of the assessors' influence is that they represented a similar perspective to BNFL. Indeed the company had based much of its evidence regarding the safe levels of emissions and individual radiation doses upon the recommendations of the International Commission on Radiological Protection (ICRP) and the National Radiological Protection Board (NRPB), both organisations of which Pochin was a senior member. BNFL consistently reiterated that they sought to keep within the standards set by those two bodies, standards Pochin had been professionally involved in approving (Stott and Taylor, 1980, pp.83-87).

The opponents openly challenged the viability of the ICRP (and NRPB) levels, arguing they were set too high. In particular the American scientist Radford, appearing as an expert witness for the opposition, argued that recent work in the USA showed the company's standards were outdated and safety levels should reflect lower levels of radiation exposure (ibid). Wynne argues that Parker saw these attacks as a challenge to the integrity of Pochin and even the Inquiry itself (Wynne, 1982, pp.116-118). To have accepted the opponents' figures would have meant rejecting the advice of the two assessors, unless they were willing to shift their professional stance. They were not, and Parker had little option but to support evidence presented by BNFL, a move that led to his castigating witnesses who cast aspersions upon the integrity of the public bodies involved in setting standards and carrying out inspection procedures (Report, 1978, various sections). This acceptance of the technical foundations for Thorp left the way clear to consider the plant on its commercial merits, a consideration Parker was also to find lay in BNFL's favour.

In many respects Parker's difficulty in accepting advice other than that interpreted for him by the technical assessors, epitomised the disunity of the engineers and scientists at the Inquiry: there was no professional opinion, only organisational goals presented in technical (and commercial etc.) terms. Radford provided an example of this. As a scientist primarily engaged upon research, he displayed a higher level of professional awareness than those from an applied background and with an organisational point to pursue. He believed that the discrepancies between his interpretation and that of Pochin could be reconciled by the two of them meeting privately and arguing through their differences (Wynne, 1982, pp.142-146). This was the attitude of the pure scientist, imbued with the sort of collegiate ethos identified by Kuhn (1970) and it clashed with the positivism of the judicial approach of Parker, an approach that recognised only 'truth' or 'untruth', not the negotiated interpretation of facts found in science (Wynne, 1982, Chapters 6, 7, 9). Wynne argues that his legalism led Parker to view the Inquiry as a pursuit of the 'truth', not to reconcile a compromise between the different interpretations of technocrats, but to choose between them (ibid).

This is not to argue that BNFL's technocrats and the other public sector witnesses were deliberately misleading in the presentation of their evidence: quite the contrary. From their organisational perspective they felt that their case was water-tight. A senior member of BNFL's Board argued argued:

> "We believed we'd got a very convincing case; there was a conscious decision to be completely open and to provide all the information that was asked for. We realised we'd got to convince and get the sympathy of the Inspector and it became very apparent very quickly that the way to do that was to be completely open with him. He was very incisive. I don't think there's anything particularly clever about it".

But because of the policies of the company it was understandable for BNFL's managers to confuse those who sought more information on THORP (or a slight delay) with groups who threatened the very existence of the company with an ideology antithetical to nuclear power.

Wynne argued (following on from Williams, 1980) that the Inquiry performed a political legitimising function through showing the decision to build THORP to be rational (1982, p.94). The need to co-opt the professional weight of the Judiciary and its rationalist ideology was due to the disharmony amongst the technocrats: he points out that 'the public legitimation and identity of nuclear power' had been deeply dependent upon the active use of scientific imagery', yet 'the strong symbolic dimension of the public role of science in this field' was severely jeopardised by the public disagreements of the technical experts (ibid). As a result the precise formalism of the legal profession was called upon to fill the gap left by the loss of public confidence in the technocrats (ibid. pp.94-97). The ideology of professionalism (as defined in Chapter 2 above) was an important element in the Inquiry process, particularly as it allowed the Judge the status to effectively arbitrate between two conflicting sets of experts, BNFL's supporters reinforcing Government policy and the aims of the corporate sector; the opposition challenging these. Wynne argued:

> "The policy process is seeking 'the right' decision, when in reality what is 'right' has already been deemed (differently) by each interest group, and the policy process is really seeking authority externally with the public at large and internally amongst the warring factions and doubters... one key factor is to have the myth of (rational) decision-making itself accepted" (ibid, p.88).

His points can be expanded to suggest that the ideology and practical social expression of professionalism allowed the higher status of the judicial process to 'rationally' arbitrate amongst the technocratic factions in favour of BNFL.

The contrast with the 'public service' ethic of the social professions (Larson, 1977) is starkly revealed again by the primary organisational loyalty of the technocrats, a loyalty of which government circles took full cognisance: a former Chief Scientist in Whitehall argued:

> "the scientists and engineers will regard themselves as employees first and as professionals second".

The organisational goals of the technocrats determined their professional stance, a phenomenon often repeated in the nuclear power policy system. The use of the Law's status and rational approach, therefore, was a political corrective to the imperfect professionalism of the technocrats; a legitimation for the wider public, convincing them that the policy of BNFL's managers corresponded with the General Welfare.

It must again be emphasised that these comments are not to suggest that the Inquiry was in any way rigged; it could not be. The policy-system waited with some trepidation to see if Parker would come to the 'decision that was expected of him' (Williams, 1980, Conclusion). One scientific advisor to the Government recalled that his Report was a 'tremendous relief' and Allday shared this collective mood, arguing happily that the whole thing had gone 'very success-fully'. Yet unless he were to reject the wisdom of his assessors, several public bodies and government departments, it is difficult to imagine how, impartial as he was, Parker could have reached any other conclusion. BNFL's case was rooted in the prevailing technical/commercial opinions of the public sector, in a sense it represented economic and technical orthodoxy. It would have been more than a little surprising if the Inquiry had overturned this. The use of technocratic reasoning, rooted in the rationalist and professional ideologies, was so entrenched at the Inquiry that personal opinions could be treated in the manner with which the DEn witness contemptuously dismissed the evidence of a journalist because the man was not a scientist (Guardian, 1978, p.54). Yet, as has been argued, such a positivistic rejection of subjective data is itself irrational in denying (at least publicly) important social concerns (Wynne, 1982, p.164). It was a deflection from the kernel of the issue: political control (ibid).

Windscale: An Appraisal

Nuclear power was locked into the corporatising system of post-war Britain (Smith, 1979); indeed, it was itself largely a product of that system (see above, Chapters 4 to 8) and previous commitments from as far back as the 1940s determined policies being followed thirty years later. It has been argued in previous parts of this study that politicians were mainly content to set broad policy parameters and then allow the technocrats to fill in the details. Where Ministers occasionally attempted to intervene on a more routine basis, there was resentment and stalling by the technocrats, mutiny usually quelled with ease. Benn argued:

> "I must say, candidly, that it is very difficult even for ministers to get all the information necessary, and this does make it hard, therefore, to retain and develop democratic control of nuclear power" (1984, p.15).

Yet such a refrain could be heard from many ministers engaged upon attempting to govern the complexities of the Welfare State: it was the nature of the beast at large, not one portion of it (cf. Crossman, 1977; Castle, 1980). A determined minister like Benn could, and did, get his way (Atom, no.244, p.23); many were content to leave the minutiae to the experts. It may well be true that politicians and civil servants were dependent upon the technocrats to explain the implications of technical policy-options, but in the case of THORP, as with the rest of the policy-system examined in this study, political/economic considerations were the domain of the politicians and if they sought to have these defined in technical terms (which suited the technocrats) then it was because they wished their chosen options to proceed without a political fuss. While there was bi-partisan and technical convergence in opinion the policy-system was usually able to have options so defined. It was 'unlucky' in the case of Windscale (and ever since) in that an educated elite no longer accepted the esoteric rationale prepared by the technocrats.

This, it can be argued, partly accounts for the dismay at Shore's decision to hold an Inquiry. BNFL certainly felt that after Benn's March announcement they had been given clearance to proceed and reports from the Cabinet had persistently led to the belief that at Windscale it was to be expansion as usual (Economist, 27/11/786). There is reason to believe that the Cabinet were particularly annoyed at the Inquiry because they had ended up in the politically embarrassing position of having secured Parliamentary approval for THORP in March as a policy statement, only to find that in establishing the Inquiry they would now be forced to have their policy scrutinised. This is one of the anomalies that led to the technocrats' attacking the policy decision-making style of Britain (Nature, v.269, p.640-41): an approach that it has been argued in several sections of this book was to misunderstand the essential nature of the policy-process in Britain, or at least call for large parts of it to be de-politicised. The opposition had served to place reprocessing (and nuclear power generally) upon the political agenda, fuelled by the growth of environmentalism, the American Moratorium on reprocessing, and a critical report by the Flowers committee in 1976. The Inquiry served notice that politicians were taking a keener interest in the details of policy, whilst attempting to maintain its technocratic definitions. Benn for one was quick to argue that the Inquiry was an opportunity to 'examine the thinking and policy-alternatives open to the Government' as well as educating 'the public' (1984, pp.20-21). Such ministerial and civil service interest is distinct, however, from parliamentary control, which remained distant.

The Windscale Inquiry illustrated once again the limits of technocratic power within the British political system and the organisational foundation of that power. The commercial strategy of BNFL, initially successful in providing greater autonomy for the company's professional/managerial elite, had a serious weakness from the professional perspective in that it opened the way for non-technical criticism, a factor not fully understood by the technocrats themselves who initially tried to dismiss it as 'subjective'. The lesson of professional disunity was also plain, a cognitive fracture the logical result of organisational loyalties, further reducing professional awareness and prefixing all technocratic advice with the knowledge (by those who received it) of its partisan bias. It can be argued, therefore, that in the field of reprocessing the technical professionals behaved more like interest groups than the definitions of Chapters 2 and 3 above would suggest; the traits inherent to ideological corporatism (Dunleavy, 1981) being submerged by this function. This merging of interest group, professional and organisational behaviour is further illustrated by the activities of the technocrats involved in other areas of the nuclear power policy-system. A particularly illuminating example is the move to secure the Joint European Torus (JET) project for Culham in Oxfordshire, where these factors became enmeshed with political questions of impressive complexity and much removed from the interests of the technocrats. With the addition of the analysis of this chapter, it is now possible to reach some tentative conclusions regarding the role of technical experts in shaping civil nuclear policy, a task attempted in the next and final chapter.

FOOTNOTES

1. Documents in the Windscale Archive (WA), University of Manchester Institute of Science and Technology (UMIST).

10 Conclusion

There have been several major arguments developed through this book regarding the role and influence of technical professionals in shaping Britain's civil nuclear energy programmes. The technocrats are an indisputably major facet of the policy system, but it is not the intention of this short final chapter merely to list the arguments of the preceding nine, rather to take them as read and briefly discuss the broad implications of the study. This book has not been concerned with the wider nuclear debate on the relative merits (or otherwise) of nuclear power per se (Sweet, 1982; Hoyle, 1977; Wynne, 1982), nor even with the task of formulating explicit recommendations for the advance of popular or political accountability (Williams, 1980; Benn, 1984), although the position of the nuclear sector within the political system has been a central theme. The focus of this study has been the singularly narrower perspective outlined in Chapter 1, namely that of engineers and scientists in the policy process.

This, however, has led to the evolution of a series of observations which relate to the issue of 'professionalism' and its interaction with organisational and political motivations. Indeed, within the nuclear power policy-system it has been argued these three factors cannot be divorced in any study of the policies inherent to it: they are inextricably intertwined and analysis which considers one without the others is in danger of a serious oversight.

First the observations can be related to the three basic propositions of Chapter 1: that professionalism forms an alternative communications system which effectively unifies 'policy development across formal agency boundaries' (Dunleavy, 1981, p.10) challenging political leadership; this leads to a corresponding reduction in the political content of decision-making in favour of professionals (Wilding, 1982); decisions finally decided upon by the policy-arbiters of the political system are those preferred by the technocrats themselves and are not, therefore, political choices. It is clear that these general propositions are an oversimplification and despite the firming-up they received in Chapter Two, cannot be said to have been substantiated by the following chapters. Aspects were certainly verified, but like the 'ideological corporatism' theory of Dunleavy (1981) and the belief by Williams that government amounted to ratification (1980, p.328), the observations can perhaps be (somewhat crudely) divided into conclusions with relevance for the study of 'professionalism', and conclusions concerned with the 'political' relationship of professional experts and policy arbiters; the two categories do, of-course, overlap at times.

Professionalism

In assessing the influence of technocrats in the process of formulating, selecting and implementing policy it became evident (through Chapters 4 to 8) that their periods of greatest control were in formulation and implementation, a factor with which the more astute technocrats themselves concurred. It must, of course, be remembered that the methodological problems inherent to this type of research preclude a verification of aspects of professional influence: nonetheless there are some arguments which can be stated tentatively.

First, the technical professions did not display the characteristics of ideological corporatism propounded by Dunleavy (1981) although at times they did provide an alternative communications system to that suggested by the formal organisational structures.To explain this point further, Dunleavy argued that where an issue becomes politicised interest corporatism declines to be replaced by ideological (or professional) corporatism (ibid, pp.7-9); this is to over-complicate the issues involved. In the nuclear power policy-system, the influence of any one group at any time relative to the other interests was constantly changing, whether it was the slow replacement of the AEA as the Government's chief nuclear adviser by the CEGB (Chapters 4 to 7) or the evolution of powerful autonomous groups (Chapter 8). Certainly there were occasions when the professional communities acted to subvert political leadership (or to replace it) via an alternative professional hierarchy (Chapter 4 on the choice of the AGR; Chapter 8 on Commercialism) however, when an issue became politicised the technocrats lobbied not as professionals, but as the representatives, indeed the employees of their employing organisation. It is the dual allegiance (Chapters 2 and 3), the merging of professional and organisational loyalty, not the former alone, that is the key to an understanding of technocratic motivation and influence. The technocrats influenced policy via their control of organisational hierarchies: if those organisations were influential then so were the technocrats, with the converse being equally applicable, as the decline in AEA status and the lack of clout wielded by THORP opponents at the Windscale Inquiry tended to illustrate (a phenomenon repeated at the Sizewell 'B' and Caithness Inquiries).

Professional power in the technological world of nuclear power is, then, circumscribed by its cognitive division into sub-disciplines and its primary manifestation is through organisational structures, the goals of which are partly informed by the professional ideology (see the technocratic moves for autonomy examined at various junctures above). The concept of 'professionalism' is dynamic, and it was suggested in Chapter 2 that the different occupations claiming professional status should be viewed within a continuum, a move that led Science and Engineering to be labelled as collections of sub-disciplines, some of which were professionalising and others which could legitimately be termed 'pseudo-professional'. It was argued that the continuum was fractured and that the move towards the status of social professions like Law and Medicine was no longer possible, if indeed it ever was (Larson, 1977).

The professional ideology itself is closely related to the ideas and motivations governing the structures of Liberal Democratic Capitalism, but the technical-professional ideology is expressly a product of the hierarchies contained within the corporate structures of this system. The rise of technocratic power paralleled (and was a result of) the rise of corporate capitalism and the demise of conflict in party politics (Smith, 1979). In post-war Britain the rise of the Welfare
State has been the rise of professional power. It is probably no accident that the increase in the politicisation of technical issues mirrors a demise in the post-war corporate consensus. This leads to a consideration of the role of politics within the nuclear policy-system, a role that defines and structures professional power. State, or more precisely, political motivations set the professional parameters: technocratic influence was exercised and sanctioned (when it was permitted to be exercised) in the interests of the wider political/economic sphere.

Political Observations

The second and third of Chapter One's general propositions, alleging technocratic power reduces the political content of decision-making and replaces it with the implementation of professional preference, missed the subtle complexities of the British political system. Wynne argued (1982), and it tended to be confirmed implicitly in Chapters 4 to 7 and explicitly in Chapter 9, that the definition of economic and social problems in technical terms disguises, but cannot eradicate, their political content. Certainly the choice of policy-options frequently represented the implementation of professional preference, but that preference and the options which represented it were formulated and defined by the broader goals of the political system. Time and again it was argued that the political desires to attain strategic energy independence, cheap electricity to fuel industry, exports of nuclear products and services, and simple nationalistic flag-waving, all set the boundaries of technocratic policy-formulation and decided which options would be successful in their selection for implementation. The technocrats best able to second-guess ministerial and civil service thinking and use that to secure political support, were those who succeeded in implementing 'their' policy choices.

In the light of this, some of Williams's conclusions need to be qualified. He argued that 'faced with matters which they could not really be expected to understand', decision-makers made their choices on an 'inadequate' basis of true knowledge' which led to government by 'ratification, indifference or bewilderment' (1980, p.328). Obviously, there was an element of this (Benn, 1984), but throughout most of its history civil nuclear power enjoyed bi-partisan support, and was not therefore a political issue in the British sense: the politicians set the general goals and within this framework the prevailing corporatist and managerial political fashions demanded that the options be defined by those best technically qualified to do so. There was little need for politicians to have a detailed knowledge: that was not their function and it is to misunderstand the nature of the political system to suggest that they should possess such a knowledge, although such a suggestion has been propounded with some vehemence by the technocrats themselves (see Chapters 4-9) in arguing against political 'interference'. The politicians had set the parameters; it was inevitable that some considerable room for manoeuvre be left to the technocrats; their socialisation, training and employee status (Chapters 2 and 3) ensured that as elements in the corporate hierarchies their options would be designed to advance the goals of that system; the options the politicians were ratifying were ones they themselves had 'ordered up' when defining the boundaries of policy. There was no real conflict over this; frictions arose over degree (such as whether to opt for gas- or water-cooled reactors (Chapters 4 to 7 above) or over the size of proposed programmes (Chapters 6 and 7) not over the fundamental goals. This 'trust' or autonomy is itself a feature of political/professional relations in society generally (Wilding, 1982). Indeed, so ingrained is this practice that the technocrats have partly evolved a misguided view of the political system, seeing the legitimate but occasional interest of politicians in the finer details of policy when an issue is politicised as interference, a perspective that fails to recognise that in Britain all things are political as and when the political sphere so defines an interest in them (Johnson, 1977).

Perhaps one of the more useful results of this study is its confirmation of the importance of what Self calls 'bureaucratic politics' (1977) and Gray and Jenkins have taken as one of the starting points in their analysis of public adminstration (1985); the importance of recognising the history, function and behaviour of organisations. More specifically, if the trap of reification is to be avoided, the motivations of senior managers in organisations and their competition for limited resources and influence has emerged as a major implication of this book. The myriad informal networks and connections within Whitehall (Heclo and Wildavsky, 1981) and between Whitehall and the periphery (Ashford, 1979; Gray and Jenkins, 1985) are increasingly well-documented. It is hoped that this study has added to that corpus. An attempt has been made to examine the close and continual links between Whitehall and quasi-governmental technocrats (Chapters 4 to 7), a relationship fraught with mutual misunderstanding, but inherent to the continuation of the policy process. Despite perennial difficulties in co-ordination and communication, the complex interactions of organisational motivation, competition and interchange provides the medium within which policy options germinate. Policy-making and administration exist, and can only exist within an organisational framework, of which the professional ideology is a part. Technical professionalism is a cognitive organisation which cuts across (and within) hierarchies of the public and private sectors. It both informs and is subservient to these hierarchies.

Future Trends

Given the admittedly narrow perspective of this study, several trends can be identified, some of which are worthy of future detailed research. The more mundane, but essential subject of professional and technical development is perhaps of the most long-term importance for the United Kingdom. It is clear that the world-view of technocrats is very different to that of Whitehall and that the managerial training of engineers and scientists should involve a preparation for the practical realities of the political implications of their work. In a sense the experience of universities and manufacturing industry since 1979 have provided an opportunity for professional (political) updating in situ. A series of studies outside of the scope of this book, however, have illustrated the woeful standard of technical development hampering Britain's economic performance (Prais and Wagner,1985; Russel, 1984; DeVille,1986; Department of Education, 1987,cmnd 9823). In order to get to grips with this problem, full cognisence should be taken of the role and influence of technologists in society and their interaction with its policy-making structures. The dearth of sociological data (see above,Chapter 3) about technocrats can only hamper desparately needed attempts to reorientate Britain's technical expertise towards the requirements of the Twenty-First Century. A closer alignment of technocratic and political perspectives upon future needs (perhaps via a flexible national science and engineering policy) would facilitate future policy-making and avoid some of the communicative problems outlined in this study.

As to the subject of civil nuclear energy; privatisation is the issue which much public debate will address. It is stated Conservative Party policy to privatise the ESI and the lessons of Chapter 8 point to a similar goal for the nuclear fuel industry. Such a move is bound to provoke much sound and fury in the environmentalist lobby, the Labour Party, and indeed within the industry itself. Following the leaks at Sellafield and the disaster at Chernobyl, a shift into the private sector for the entire nuclear industry will not prove easy. The consciousness of those in the private sector and their attitude towards such things as safety standards and completion dates is bound to be different to that adopted by public servants or those more closely aligned to a professional code of ethics, no matter how cursory their enforcement. A candid example was provided by a senior BNFL manager who pointed-out that the introduction of new NII siesmic standards had led to a large increase in costs at Sellafield and new power stations under construction. At the time these were introduced, along with new safety standards, their impact was not fully appreciated and the manager argued that, "I have a feeling that if we were in the private sector it might be that the policy of the NII might have been challenged much earlier." Furthermore, the profit motive would have ensured that the cost of the NII's decision would have been identified earlier than it was. It is often difficult to reconcile short-term profit and political motives with longer term public benefits, a problem which will be exacerbated by privatisation; strategic energy planning and public confidence in 'experts' will be victims of such a manoeuvre.

PWRs will also remain a subject of much opprobrium for as long as they remain on the agenda. Oppenshaw states that in his view the American nuclear manufacturing industry intends to survive the current depression by supplying "their obsolete plant to countries stupid enough to think they are a good buy" (1986, p.320). There is no reason to believe the anodyne conclusions of Layfield did anything to allay the deep fears of those who hold this and stronger views with regard to nuclear power in general and the PWR in particular. To dismiss them as anti-industry, as the ESI has done in the past, is both simplistic and counter-productive, provoking greater polarity where consensus is the only sensible course for future progress.

It is clear that technocrats have indeed influenced nuclear policy and will continue to do so, on terms laid down by the policy arbiters: and this is the nub of nuclear politics. The grounds for debate and the opportunities for participation are strictly limited and jealously protected by a small group of politicians and civil servants. This is not to slip into cosy elite and conspiracy theories, but to reach a rather pessimistic conclusion. The public circus of the Sizewell Inquiry, like the one at Windscale before it, added not one jot to genuine participation in the policy making process by the citizenry or by technocrats ill-disposed towards official proposals. Even this sadly neutered version of popular involvement seems set never to be repeated. Influence upon nuclear, as upon any other policy, therefore remains restricted, controlled, and essentially secretive, and that is the lesson of this policy-study into just one aspect of the Government of Britain.

References and bibliography

Official Documents (all London, HMSO)

The Nuclear Power Programme, Cmnd.1083, 1960.
The Second Nuclear Power Programme, Cmnd.2335, 1964.
Nuclear Power Policy H.C.350, Cmnd.5499, 1973.
Choice of Thermal Reactor Systems, Cmnd.5731, 1974.
Royal Commission on Environmental Pollution, 6th Report., Cmnd.6388, 1976.
Energy Policy, Cmnd.7101, 1978.

Select Committee on Science and Technology (and Select Committee on Energy)

"The United Kingdom Nuclear Reactor Programme." HC.381-xvii (1966-7).
"Nuclear Power Policy", HC.350 (1972-3), HC.444 (1971-2), HC.117 (1972-3).
"The Choice of a Reactor System", HC.145 (1973-4), HC.73.
"The SGHWR Programme", HC.89 (1976-7), HC.623 (1975-6).
1st Report HC.114-1 114-11 (1981) Vols.I-IV.
"Nuclear Power" Cmnd.8317. (Government Reply).
"The Windscale Inquiry", Report by the Hon.Mr. Justice Parker, 1978.
"Dungeness B AGR Nuclear Power Station", (CEGB, London, 1965).
"Energy Policy REview", DEn, Energy Paper no.22, 1977.
"Atomic Energy in Britain", UKAEA, London, 1979.
"Electricity Supply in the United Kingdom", CEGB, London, 1980.
"Thermal Reactor Strategy", CEGB, London, 1981.
Evidence to the Sizewell 'B' Inquiry, CEGB, London, 1981.

Parliamentary Debates and Statements

Lords
13/4/67, Fuel and Energy Policy.
8/5/68, The Nuclear Power Industry
14/4/70, Atomic Energy Authority Bill
23/4/70,   "
26/1/71   "
4/2/71   "

Commons
18/11/59, Atomic Energy Authority Bill.
3/2/61, Nuclear Power Programme.
19/11/62, Atomic Energy Establishment, Capenhurst.
25/5/65, Nuclear Power (Dungeness B Station).
20/3/67, Nuclear Power Stations.
17/12/70, Atomic Energy Authority Bill
15/1/71, Atomic Energy Authority Bill.
8/8/72, Nuclear Reactor Policy.
20/12/74, Nuclear Energy.
16/3/76, JET.
13/12/76, Windscale.
20/12/76, Nuclear Power;.
21/3/78, Windscale Inquiry Report.
15/5/78, Windscale (Special Development Order).

Annual Reports of:
The United Kingdom Atomic Energy Authority, 1956-7 to 1982-3.
The Central Electricity Generating Board, 1960-1 to 1982-3.
British Nuclear Fuels Limited.
Joint European Torus.

European Documentation
5th to the 14th General Reports of the European Communities, London.
Bulletins of the European Communities (1969-1980).
Communications of the European Commission to the Council of Ministers (75)350; (76)8; (76)271; (76)523; (78)147; (78)158, London.

Journals - Between 1958 and 1983, various issues
Atom; The Economist; The New Scientist; The New Statesman; Nature; Nuclear Power; Journal of the Institution of Nuclear Engineers; Nuclear Engineering International; Journal of the British Nuclear Energy Society; Nuclear News; Nuclear Fusion; Contemporary Physics; Atomic Energy Review.

Conference Papers and Speeches etc.
"Implications of Radioactive Waste", EPEA, Chertsey, 1978.
"Progress and Prospects of Nuclear Fusion", R. Pease, 1981 to BNES.

"Nuclear Power", C. Allday, 1978, to BNES.
"Issues in the Sizewell 'B' Inquiry", Six volumes. Centre for Energy Studies, Polytechnic of the South Bank, London 1982.
"Nuclear Power", A. Palmer, EPEA, London, 1983.
(Also the Windscale Archive, UMIST, Manchester).

Engineering and Science Reports; evidence on engineering etc.
Monopolies Commission, Cmnd.4463, 1970.
"Engineering Our Future", M. Finniston, Committee of Inquiry into the Engineering Profession", Cmnd.7749, 1981.
The Surveys of Professional Engineers (First Ministry of Technology and CEI, then CEI alone, after 1971).
The Surveys of Professional Scientists (MinTech and CSI).
"The Organisation as Bias of Professional Status". H.B. Watson, PhD thesis, LSE, 1975.
Engineering Industry Training Board Salary and Location Surveys.

"Professional and Managerial Staff in Engineering", Tass, Hounslow, 1979.
"Journal" of ASTMS, May-August 1981.
EPEA Annual Report 1974.
Evidence to Finniston Committee from the Institution of Chemical Engineers.
Submission to the Lords Select Committee on Science and Technology, 1982, by ASTMS.

BIBLIOGRAPHY

Professionalism, Engineering and Science
Banting, K.G. (1979) Poverty, Politics and Policy, London, Macmillan.
Bell, D. (1974) The Coming of the Post-Industrial Society, London, Heinemann.
Berthaud, R. and Smith, D. (1980) The Education, Training and Careers of Professional Engineers, London, HMSO.
Clark, N. (1985) The Political Economy of Science and Technology, Oxford, Basil Blackwell.
Cockburn, C. (1977) The Local State, London, Pluto.
CEI, (1975) Professional Engineers and Trade Unions, London, CEI.
Daniels, A. (1975) "How Free should Professions Be?" in E. Friedson (ed) The Professions and their Prospects, Beverley Hills, Sage.
Dennis, N. (1970) People and Planning, London, Faber.
Dingwall, R. and Lewis, P. (1983), The Sociology of the Professions, London, Basingstoke, Macmillan.
Dunleavy, P. (1979) The Policy Implications of Professionalism; Paper for PSA, ILGS, Birmingham.
Dunleavy, P. (1980) Urban Political Analysis, London, Macmillan.
Dunleavy, P. (1981) "Professions and Policy Change" in Public Administration Bulletin, no.36, pp.3-16.
Elliot, P. (1972) The Sociology of the Professions, London, Macmillan.
Engineering Council, (1985) Survey of Chartered and Technician Engineers, London, Engineering Council.
Etzioni, A. (1969) The Senior Professions and their Organisation, New York, Free Press.

Fay, B. (1975) Social Theory and Political Practice, London, Allen and Unwin.
Friedson, E. (1970) Professional Dominance, New York, Atherton.
Friedson, E. (ed) 1973) The Professions and their Prospects, New York, Sage.
Gerstyle, J. and Hutton, S.P. (1966) Engineers: The Anatomy of a Profession, London, Tavistock.
Gerstyle, J. and Jacobs, G. (eds) 1976) Professions for the People, New York, Scherliman.
Goode, W. (1966) "Professions and Non-Professions" in Volmer and Mills, Professionalization, New York, Prentice Hall.
Greenwood, E. (1965) "Attributes of a Profession" in M. Zald (ed) Social Welfare Institutions, London, Wiley.
Habbermas, J. (1979) Communication and the Evolution of Society, London, Heinemann.
Howie, W. (1977) Trade Unions and the Professional Engineer, London, Telford.
Illich, I. (1973) Deschooling Society, London, Penguin.
Illich, I. (1977) Disabling Professions, London, Boyars.
Jackson, J.A. (1970) Professions and Professionalisation, Cambridge University Press.
Johnson, T. (1972) Professions and Power, London, Macmillan.
Kuhn, T. (1970) The Structure of Scientific Revolutions, Chicago, Chicago University Press.
Larson, M. (1977) The Rise of Professionalism, University of California Press.
Littlejohn G. et al. (1978) Power and the State, London, Croom Helm.
Mannheim, K. (1954) Ideology and Utopia, London, Routledge and Kegan Paul.
Massey, A. (1985) The Joint European Torus, paper presented to ECPR Conference, Barcelone, April.
Millerson, G. (1964) The Qualifying Associations: A Study in Professionalisation, London, Routledge and Kegan Paul.
Parsons, T. (1967) The Social System, London, Routledge and Kegan Paul.
Price, D. (1965) The Scientific Estate, Oxford University Press.
Vollmer, H.M. and Mills, D.L. (1966) Professionalisation, New York, Prentice Hall.
Wilding, P. (1982) Professional Power and Social Welfare, London, Routledge and Kegan Paul.

Nuclear Energy

Benn, T. (1984) The Sizewell Syndrome, Nottingham, Spokesman.
Burn, D. (1978, reprinted 1980) Nuclear Power and the Energy Crisis, London, Macmillan.
Cockcroft, JH. (1950) The Development and Future of Nuclear Energy, Oxford, Clarendon Press.
Evans, N. and Hope, C. (1984) Nuclear Power, Cambridge University.
Fernie, J. (1980) A Geography of Energy in the United Kingdom, London, Longman.
Forman, N. Towards a More Conservative Energy Policy, London, Conservative Political Centre.
Gowing, M. (1974) Britain and Atomic Energy, London, Macmillan, (2 vols.)

Greenhalgh, G. (1980) The Necessity for Nuclear Power, Graham and Trotter.
Gummett, P. (1980) Scientists in Whitehall, Manchester University Press.
Hannah, L. (1982) Engineers, Managers and Politicians, London, Macmillan.
Hinton, C. (1958) "Nuclear Power Development", J.Inst.Fuel, March 1958
Hoyle, F. (1979, 2nd edn.) Energy or Extinction, London, Heinemann.
Ince, M. (1982) Energy Policy, London, Junction Books.
Lovins, A. (1977) Soft Energy Paths, London, Penguin; London, FoE.
Marsh, N.F. (1967) The Work of Engineers and Scientists in the Electricity Supply Industry, London, The Electricity Council.
Marshal, W. (ed) (1983) Nuclear Power Technology, London, Clarendon.
Openshaw, S. (1986) Nuclear Power: Siting and Safety, London, Routledge and Kegan Paul.
Patterson, W. (1976) Nuclear Power, London, Penguin.
Pearson, L. (1981) The Organisation of the Energy Industry, London, Macmillan.
Sweet, C. (ed) (1980) The Fast Breeder Reactor
Sweet, C. (1982) The Costs of Nuclear Power, London, Anti-Nuclear Campaign.
Williams, R. (1980) The Nuclear Power Decisions, London, Croom Helm.
Willson, D. (1981) A European Experiment, Bristol, Adam Holger.
Wynne, B. (1982) Rationality and Ritual, BSHS Monograph.

see also:
Dickinson, D. (1974) Alternative Technology, London, Fontana.
Schumacher, E. (1974) Small is Beautiful, London, Abacus.
Vig, N. (1968) Science and Technology in British Politics, London, Pergamon.
Wearf, S. (1979) Scientists in Power, Harvard.

Public Administration/Policy (where not included in the above)
Albrow, M. (1970) Bureaucracy, London, Macmillan.
Ashford, D. (1979) Policy and Politics in Britain, London, Basil Blackwell.
Bacharach, S. and Lawler, E. (1980) Power and Politics in Organisation, San Francisco, Josey Bass.
Benn, T. (1981) Arguments for Democracy, London, Macmillan.
Bourn, J. (1974) The Administrative Process as a Decision-Making and Goal Attaining System, Milton Keynes, Open University.
Brown, R. and Steel, D. (1979) The Administrative Process in Britain, London, Methuen.
Burch, M. and Wood, B. (1983) Public Policy in Britain, Oxford, Martin Robertson.
Crossman, R. (1977) Diaries of a Cabinet Minister, London, Hamilton and Cape.
Downs, A. (1967) Inside Bureaucracy, Boston, Little Brown.
Dunleavy, P. (1982) "Is there a Radical Approach to Public Administration?" in Public Administration 60, 215-224.
Ellul and Boguslaw (1981) The New Utopians, Lovington.
Gray, A. and Jenkins, W. (1985) Administrative Politics in British Government, Brighton, Wheatsheaf.

Greenwood, J. and Wilson, D. (1984) Public Administration in Britain, London, Allen and Unwin.
Ham, C. and Hill, M.J. (1984) The Policy Process in the Modern Capitalist State, Brighton, Wheatsheaf.
Heald, D. (1983) Public Expenditure, Oxford, Martin Robertson.
Heclo, H. and Wildavsky, A. (1981) The Private Government of Public Money, London, Macmillan.
Johnson, N. (1977) In Search of the Constitution, London, Macmillan.
Jones, G. (ed) (1980) New Approaches to the Study of Central Local Relations, Aldershot, Gower.
Kellner, P. and Crowther-Hunt, Lord (1980) The Civil Servants, London, Macdonald.
Likierman, A. (1982) Management Information Systems foe Ministers, Public Administration, Vol 6, pp. 127-142.
Lindblom, C. (1965) "The Science of Muddling Through", Public Administration Review, 19, 79-88.
Niskanen, W.A. (1971) Bureaucracy and Representative Government, Chicago, Atherton.
O'Leary, B. (1985) "Is there a Radical Public Administration?" Public Administration 63, pp.345-351.
Peters, G. (1978) The Politics of Bureaucracy, New York and London, Longman.
Pfeffer, J. (1981) Power in Organisations, Marchfield, Pitman.
Pitt, D. and Smith, B. (1981) Government Departments, London, Routledge and Kegan Paul.
Pitt, D. and Smith, B. (eds) (1984) The Computer Revolution in Public Administration, Brighton, Wheatsheaf.
Saunders, P. (1983) The Regional State, Paper presented at Nuffield College, Oxford, September.
Self, P. (1977) Administrative Theories and Politics (2nd ed.) London, Allen and Unwin.
Self, P. (1977) Econocrats and the Policy Process, London, Allen and Unwin.
Smith, T. (1979)The Politics of the Corporate Economy, London, Martin Robertson.
Stanyer, J. and Smith, B. (1976) Administering Britain, London, Fontana.
Vickers, G. (1965) The Art of Judgement, London, Chapman and Hall.
Wass, D. (1984) Government and the Governed, London, Allen and Unwin.
Weber, M. (1947) The Theory of Social and Economic Organisation, New York, Free Press.